Creativity in Teaching and Teaching for Creativity

In this book, the authors write about creativity in teaching and how to enhance creativity in learners. They highlight the new reality of teaching and learning in the digital era, specifically the impact of artificial intelligence, data economy, and artificial minds on modern teaching practices, curriculum design, and the role of teachers in classrooms.

Creativity in Teaching and Teaching for Creativity: Modern Practices in the Digital Era approaches human intelligence as a universal gift. It emphasizes that the creativity of human beings is not only a natural quality, but one that can be enhanced as a result of learning. This book suggests new teaching models and approaches and discusses how the role of teachers in the classroom has fundamentally changed, emphasizing the emotional connection between students and teachers.

The book will also find interest among higher education policymakers who believe in the transformation of the education industry, research scholars pursuing their Ph.D. in the fields of education technology and education and learning, as well as industry professionals in education management and e-learning companies working in the areas of education technology and artificial intelligence.

Creativity in Teaching and Teaching for Creativity
Modern Practices in the Digital Era in Engineering

Lucy Lunevich and Majed Wadaani

CRC Press is an imprint of the
Taylor & Francis Group, an **informa** business

First edition published 2023
by CRC Press
6000 Broken Sound Parkway NW, Suite 300, Boca Raton, FL 33487-2742

and by CRC Press
4 Park Square, Milton Park, Abingdon, Oxon, OX14 4RN

CRC Press is an imprint of Taylor & Francis Group, LLC

© 2023 Taylor & Francis Group, LLC

Reasonable efforts have been made to publish reliable data and information, but the author and publisher cannot assume responsibility for the validity of all materials or the consequences of their use. The authors and publishers have attempted to trace the copyright holders of all material reproduced in this publication and apologize to copyright holders if permission to publish in this form has not been obtained. If any copyright material has not been acknowledged please write and let us know so we may rectify in any future reprint.

Except as permitted under U.S. Copyright Law, no part of this book may be reprinted, reproduced, transmitted, or utilized in any form by any electronic, mechanical, or other means, now known or hereafter invented, including photocopying, microfilming, and recording, or in any information storage or retrieval system, without written permission from the publishers.

For permission to photocopy or use material electronically from this work, access www.copyright.com or contact the Copyright Clearance Center, Inc. (CCC), 222 Rosewood Drive, Danvers, MA 01923, 978-750-8400. For works that are not available on CCC please contact mpkbookspermissions@tandf.co.uk

Trademark notice: Product or corporate names may be trademarks or registered trademarks and are used only for identification and explanation without intent to infringe.

Library of Congress Cataloging-in-Publication Data
A catalog record has been requested for this book.

ISBN: 978-1-032-35824-6 (hbk)
ISBN: 978-1-032-35878-9 (pbk)
ISBN: 978-1-003-32914-5 (ebk)

DOI: 10.1201/9781003329145

Typeset in Times
by codeMantra

Contents

Preface — vii
Contributing Authors — ix
Author Biographies — xi

1 Human Creativity and Giftedness — 1
Lucy Lunevich
　1.1　Introduction — 1
　1.2　Creative Destruction — 2
　1.3　Creativity and Giftedness — 4
　References — 5

2 Human Intelligence and Giftedness — 7
Lucy Lunevich
　2.1　Introduction — 7
　2.2　Human Intelligence and Artificial Intelligence — 7
　2.3　Human Intelligence and Giftedness — 9
　References — 11

3 Critical Digital Pedagogy – Innovative Model — 13
Lucy Lunevich
　3.1　Introduction — 13
　3.2　Re-Emerging Pedagogy — 15
　3.3　Innovative Model for the Digital Era — 19
　3.4　Digital Learning — 21
　3.5　Conclusion — 23
　References — 24

4 Creativity in Teaching and Teaching for Creativity – Engineering Students — 29
Lucy Lunevich
　4.1　Introduction — 29
　4.2　Creative Pedagogy — 30
　4.3　Art in Teaching — 32
　4.4　Teaching for Creativity — 33

	4.5	Critical Thinking and Problem-Solving Skills	34
	4.6	Observation in Education Research	35
	4.7	Involvement of Students in Education Enquiry	36
	4.8	Research Methodology	37
	4.9	Ethics of Observation	38
	4.10	Results	38
	4.11	Conclusion	40
	References		41

5 Creativity and Human Development towards Self-Actualization   45
Majed Wadaani

	5.1	Introduction	45
	5.2	Creativity as a Human Phenomenon	46
	5.3	Creativity and Human Intelligence	51
	5.4	Creativity as Human Development towards Self-Actualization	53
	5.5	Teaching for Creativity	55
	5.6	Conclusion	59
	References		59

6 Supporting Creativity and Mathematical Talent Development   63
Majed Wadaani

	6.1	Introduction	63
	6.2	Research Problem and Questions	65
	6.3	Significance	66
	6.4	Method and Procedures	66
	6.5	Review of Literature	67
	6.6	Results and Discussion	71
	6.7	Conclusion	76
	References		77

Preface

Creativity in Teaching and Teaching for Creativity: Modern Practices in the Digital Era in Engineering has been written to assist teachers and educators around the world to deeply engage in their critical mission of teaching. It presents the observational research conducted with engineering students over four years. Digital literacy skills are known as the 4Cs of 21st-century skills: critical thinking, communication, collaboration, and especially creativity, have been a major focus in this book.

We have been blessed to spend some time thinking and writing the book, as we sincerely believe the materials presented in this book will help millions to improve their life as a result of learning from teachers who understand creativity in teaching and teaching for creativity. In educational contexts, the three components, i.e. technology, pedagogy, and learning content, mutually interact with one another, and the interaction creates a new learning environment. In particular, we believe that certain school beliefs and practices are inconsistent with the kinds of findings that have emerged from our research. These beliefs and practices should be changed.

First, students should not be assigned to tracks, sections, or other instruction conditions on the basis of a single IQ score or even two scores obtained from conventional tests of intelligence. The abilities that can be brought to bear on school work are broader than what conventional tests measure.

Second, instruction and assessment should reflect this broader range of abilities. In particular, many have suggested that analytical, creative, and practical abilities serve as a basis for identifying abilities, providing instructions, and administering assessments.

Third, teachers should allow flexibility in placements, recognizing abilities are not fixed but developing and that students with the same level of abilities on a static test may benefit differentially from instructions, as would be revealed by a dynamic test.

Finally, teachers the curriculum design should recognize that students may come from backgrounds where what is valued differs from teachers' value. Students will learn better if teachers understand and take students' values into account, showing an amazing process of learning.

Contributing Authors

Lucy Lunevich, Ph.D.,
RMIT University, Australia.

Majed Wadaani, Ph.D.,
Jazan University, Saudi Arabia.

Author Biographies

Prof. Lucy Lunevich is a Senior Lecturer and Program Director of the Growth Lab at RMIT University, School of Engineering, Melbourne, Australia. She is a passionate advocate for the role of higher education, research, and innovation in creating economic prosperity and for improving people's lives through education and training, innovation, and environmental awareness. She maintains an active social media presence and has been rated as one of the 25 smartest women to follow on Twitter by *Fast Company* magazine. She is the chief editor of the Water Environment Federation (WEF) publishing agency in the US and is frequently invited to address businesses in Australia and the US. Prof. Lunevich's research interests include areas of critical digital pedagogy and meta-pedagogy as emerging fields of research and practice.

Dr. Majed Wadaani is an Associate Professor and Chairman in the Department of Special Education, College of Education, Jazan University, Saudi Arabia. He has been working in the areas of thinking skills, creativity development and self-actualization, gifted education, and student support to demonstrate their full potential.

Human Creativity and Giftedness

Lucy Lunevich

1.1 INTRODUCTION

Theories of human creativity do not all theorize the same thing; rather, they tend to theorize different aspects of creativity. Moreover, the scholars proposing these theories rarely clarify which aspects of creativity their theories embrace. Consequently, it is difficult to know which aspects of the various theories are complementary and antagonistic, and even harder to apply them in design curriculum at university or school education.

Human creativity refers to the novelty in anything. Creativity means different things to different people, cultures, civilizations, and religions. Plato, Kant, and Vygotsky define creativity as a state of mind that one experiences platonic love, inspiration, renewal, growth, self-discovery, higher meaning of life, stillness, connection to surroundings, and goodness. In terms of Noahide theology, the soul or spirituality within humans is the faculty or capacity for reception of resonance with and recognition of the Divine. When this capacity is activated as a "functioning" soul or conscience, it orients the mind to the Divine, higher creativity (Cowen, 2014). While intellect is the distinguishing hallmark of humans, intellect itself is placed in an equivocal position (Cowen, 2014). It can produce an elaborate defence or rationalization of mere predispositions, many of which arise from the corporeal-emotional personality of the human (Cowen, 2014). It could regulate emotion according to principles beyond simply willed predisposition and predilection (Cowen, 2014). It is explained that the biblical statement that the human was made in the image of God refers to intellect, creativity orientated upwards towards higher knowledge.

1.2 CREATIVE DESTRUCTION

Human civilization has experienced sparks of creativity at different historic times to solve problems in new ways. It can be seen as failure and demolition of the old concepts, structures, economic and social orders – destruction and the same time emerging new economic and social orders through creativity. Creativity intertwined in the forms of destruction is well understood in the art world.

In fact, creativity is inseparable from destruction. Rabbi David Aaron (2009) said we often fail at creating intimate loving relationships because we are dominated by our creative side. He pointed out that our creative side searching for godly dignity is success-minded; therefore, we enter into relationships because we understand that two heads are better than one (Aaron, 2009). A clever entrepreneur knows that pooling talents and resources together increases the chance of succeeding in our conquest of Earth. Together we may achieve dignity but not spiritual completion.

Aaron (2009) said that to achieve true love and intimacy, we need to nurture our sacred side, which is the opposite of a creative one. We must turn our attention towards living a complete relationship. If a teacher is asked by their students, "Why do you love teaching us?", we might want to hear: "No reason, just because our planet needs to reclaim its beauty and love through all of us, living creatures, economic and social development is part of it, not against it". If there is a reason, it is not teaching with the platonic love of students.

Sternberg (2001) referred to creativity as the potential to produce novel ideas and of high quality (Sternberg, 1998, 2001, 2010). He continued that creativity in society is best understood in terms of a dialectical relation to intelligence, wisdom, and giftedness.

Sternberg (2001) pointed out that definitions of creativity, like definitions of intelligence, differ. Still, they have in common their emphasis on people's ability to produce products and resolve problems in creative ways that are high in quality and also novel, unique (Sternberg, 2001; Sternberg et al. 1997). Many highly creative individuals "defy the crowd" (Sternberg & Lubart, 1991, 1996) by showing new ways of doing things, finding ways to cut through barriers and difficulties others see as unbearable (Lunevich,

2021). This view implies that creativity is always a person-system interaction. When the system changes, creativity sparks as a result of new possibilities. It seems creativity is meaningful only in the context of a system that judges it, and what is creative in one context may not be in another (Lunevich, 2021; Sternberg & Lubart, 1996). For example, there was a time in French history (-1770–1790) when designers had to produce new collections each week to stay in business and promote their work to be acknowledged by the king. How is this possible? Where does creativity come from?

Hence, creativity must be viewed as a property of an individual as that individual interacts with one or more systems, learning about the state of the system and trying to improve it. For example, Leonardo studied the flight of birds with his customary zeal, and it is well known that he made designs for a flying machine. Most of Leonardo da Vinci's aeronautical designs were ornithopters, machines that employed flapping wings to generate lift and propulsion. He sketched such flying machines with the pilot prone, standing vertically, using arms and legs. He drew detailed sketches of flapping-wing mechanisms and means for actuating them, learning from nature, testing, and changing models as the result of exhaustive observation of the movements of birds to examine how the wings achieved flight.

Another example is Miuccia Prada. Prada changed the way the world saw fashion. Among other things, opting for not a single look identifying her brand but ricocheting from one idea to another, and another, and another, every season. This forced the rest of the world to run to keep up with her. She demonstrated an instinct for what was right at a certain moment that was second to none. It is clear that Prada is well educated and knowledgeable about art, film, music, and culture, present and past (Frankel, 2019). Her references are layered and complex. Still, there is an open-mindedness, curiosity, and warmth to her that is equally important. It is felt in her physical presence and the bravery of everything, from her words to her collaborations, and from the films, exhibitions, books, and retail environment she has developed to the merchandise displayed therein (Frankel, 2019). She is hugely courageous, and her output is intensely personal.

Among other things, highly creative people decide to redefine problems, frame them in different ways, so the problem itself becomes a solution. They analyse their ideas, attempt to persuade others of their value, dream with their ideas, and take sensible risks. Sternberg (1996, 2001, 2006) pointed out that an implication of this view of creativity as a decision is that anyone can adopt a creative attitude and think creatively. He continued that, for various reasons, people will not typically reach the heights of creativity of the individuals whose contributions are reviewed in this chapter. Among these reasons are different degrees of compatibility between people's thinking and where a field is at a given time in history. For example, someone spontaneously generating the ideas underlying impressionism today will be understood at other times of history and interpreted differently by artificial intelligence.

Creative people often feel underappreciated and confused with their ideas. This is normal because the idea requires translation into new forms, new meanings, new approaches, and new business models, for example.

This is the required giftedness of those who are creative.

1.3 CREATIVITY AND GIFTEDNESS

Creative people have a highly intuitive way of working. They create a fun environment to work with, but one that is also very demanding for all. That is why they are open and at ease while working. They talk, interact in many ways, and energize each other in this creative, demanding environment, where it is easy to burn out, catch a fire of creativity and start again. There is always that sense of being on the edge of making sure what you do makes sense.

Translation of creative ideas and novel ideas into the material things is giftedness. "The moment you start being in love with what you are doing and thinking it's beautiful or rich, then you are in danger", Miuccia Prada told Cathy Horyn of the *New York Times* before the 2001 collection was shown (Frankel, 2019). "You have to always work against what you did before and even against your own taste". Wow! Is it true?

Sternberg (1991, 2000) described that gifted individuals whose contributions have been historically enduring had demonstrated three common attributes that can be revealed while analysing their contributions. According to Sternberg (1985, 1996, 2000), these include:

- analytical: the ability to analyse and evaluate one's own ideas and those of others
- creativity: the ability to generate one or more major ideas that are novel and of high quality
- practical: the ability to convince people of the value of ideas that are novel and of high quality.

In fact, the practical aspect of creativity could be tough and enhanced for all creative people; students consider creativity to be tough in schools, universities, and other educational bodies. So, creative people could be further developed with the giftedness of analytical and practical skills.

The kind of tasks people face in everyday life often requires all three types of thinking – creative, practical, and analytical abilities. The creator might display very high levels of creative ability but not analytical or practical ability. As we know, these individuals generate ideas fairly easily but then cannot fully analyse these ideas or put them into practice (Sternberg, 2001; Sternberg et al., 2008). Most of us experience creators at some point, but realization of ideas is frequently the hardest part. If they lack the analytical ability needed to shape an idea, to rid of its weak elements and develop its strong elements, they will be unable to follow through on the idea (Sternberg, 2000, 2004). They may also lack the practical ability to convince people of the value of their idea or find others to help shape their ideas (Stenberg, 1985, 1991, 2001). According to Sternberg (2000, 2004), the practitioner is an individual with highly developed practical skills without complementary analytical and creative skills. These individuals are persuasive and often entertaining but lack substance in their thinking (Sternberg, 1991, 2000; Sternberg & Kaufman, 1998).

The analyst with high analytical abilities but not other types of abilities fare well in many academic environments. However, such individuals may be less likely to make a creative contribution as adults than individuals with other patterns of giftedness because they lack creative, productive abilities. This might be partly due to the education system they were exposed to, which does not support creativity and learning for creativity. Curriculum design might be dry and boring, frequently dominated by technical subjects, focusing on rational aspects of learning instead of creating a pleasant social environment and student's emotional involvement in each step of learning (Lunevich, 2021).

Some believe that creativity and practicality are two different mindsets, different ways of thinking. Others argue that the two different mindsets conflict. However, both can be taught in the classroom so that students can reach their full potential and self-actualization, as discussed in this book, specifically in Chapter 5.

REFERENCES

Aaron, R. D. (2009). *The God-Powered Life.* Boston, MA & London: Trumpeter.
Cowen, S. D. (2014). *The Theory & Practice of Universal Ethics. The Noahide Laws.* Melbourne & New York: The Institute for Judaism and Civilization, Inc.
Frankel, S. (2019). *Prada.* London: Thames & Hudson Ltd.
Lunevich, L. (2021). Creativity in Teaching and Teaching for Creativity in Engineering and Sciense in High Education – Revisiting Vygotsky's The Psychology of Art. *Research Journal of Creative Education, 12,* 1445–1457.
Sternberg, R. (1985). *Beyond IQ: A Triarchic Theory of Human Intelligence.* New York: Cambridge University Press.

Sternberg, R. (1991). An Investment Theory of Creativity and Its Development. *Human Development, 34*, 1–34.
Sternberg, R. (1998). Human Abilities. *Annual Review of Psochology, 49*, 479–502.
Sternberg, R. (2000, June). Patterns of Giftedness: A Triarchic Analysis. *Roeper Review*, 231.
Sternberg, R. (2001). What Is the Common Tread of Craetivity? Its Dialectical Relation to Intelligence and Wisdom. *American Psychologist, 56*(4), 360–362.
Sternberg, R. (2004). Teaching College Students that Creativity Is a Decision. *Guidance & Counseling, 19*, 196–200.
Sternberg, R. (2006). The Nature of Creativity. *Creativity Research Journal, 5*, 394–414.
Sternberg, R. (2010). Teaching for Creativity. In R. Beghetto, & J. Kaufman (Eds.), *Nurturing Creativity in the Classroom* (pp. 394–414). New York: Cambridge University Press.
Sternberg, R. K. (2008). *Applied Intelligence.* New York: Cambridge University Press.
Sternberg, R., & Kaufman, J. (1998). Human Abilities. *Annual Review Psychology, 49*, 47–502.
Sternberg, R., Kaufman, J., & Grigorenko, E. (2008). *Applied Intelligence.* New York: Cambridge University Press.
Sternberg, R., & Lubart, T. (1996). Investing in Creativity. *American Psychologist, 51*(7), 677–688.
Sternberg, R., O'Hara, L., & Lubart, T. (1997). Creativity as Investment. *California Management Review, 40*(1), 8–21

Human Intelligence and Giftedness

2

Lucy Lunevich

2.1 INTRODUCTION

Human intelligence is considered to be the human mind's ability to connect intellectual ability with heart. Sternberg (1997, 1998, 2008) and others define intelligent people as those who somehow acquire the skills that lead them to fit into existing environments. Generally, intelligent people are rewarded precisely because they adapt, often in multiple environments. For instance, the same person may be successful in different environments like business, academia, and government.

2.2 HUMAN INTELLIGENCE AND ARTIFICIAL INTELLIGENCE

Significant technological progress allows us to use artificial intelligence and have access to real-time information. Some compare human intelligence with

DOI: 10.1201/9781003329145-2

artificial intelligence. This should have never happened. Artificial intelligence is an artificial intellect, but not real intelligence itself, because AI does not have emotions, soul, love. It does not suffer or cry as a human can and will as a result of being intelligent.

It is expected that human intelligence is the intellectual capability of humans, which is marked by complex cognitive feats and high levels of motivation and self-awareness. High intelligence is associated with better outcomes in life. Some define human intelligence as a mental quality that consists of the abilities to learn from experience, adapt to new situations, understand and handle abstract concepts, and use knowledge to manipulate one's environment. Human intelligence should never be compared to AI because human intelligence has a spiritual element to it. Ken O'Donnell advocates the integration of spiritual intelligence (SQ) with both rational intelligence (IQ) and emotional intelligence (EQ). IQ helps us to interact with numbers, formulas, and things; EQ helps us to interact with people; and SQ helps us to maintain inner balance. To calculate one's level of SQ, he suggests the following criteria:

- How much time, money, energy, and thought are needed to obtain a desired result?
- How much bilateral respect exists in our relationships?
- How "clean" a game we play with others?
- How much dignity we retain in respecting the dignity of others?
- How tranquil we remain despite the workload?
- How sensible our decisions are?
- How stable we remain in upsetting situations?
- How easily we see virtues in others instead of defects?

Robert Emmons (2000) defines SQ as "the adaptive use of spiritual information to facilitate everyday problem solving and goal attainment". He originally proposed five components of SQ (Emmons, 2000):

1. The capacity to transcend the physical and material.
2. The ability to experience heightened states of consciousness.
3. The ability to sanctify everyday experience.
4. The ability to utilize spiritual resources to solve problems.
5. The capacity to be virtuous.

The higher level of SQ is self-awareness. To calculate one's level of human intelligence, all criteria need to be considered (Emmons, 2000).

The question is what is the role of pedagogy? How much learning do humans receive from structural enrolment places like schools and universities, and how much learning from an open environment – social, family, interactions with friends, and street education? Many agree that to be successful in life, one needs both a university education and "street education", particularly the ability to read the environment, get along with people, and manage yourself. This is why creating a supportive social environment in the classroom is critical for transformative learning (Emmons, 2000).

2.3 HUMAN INTELLIGENCE AND GIFTEDNESS

Transformative learning could produce more gifted students. This theory is also supported by Vygotsky's research, which stated that teachers should lead learners without preconditions of students' abilities (Vygotsky, 2016).

Transformative learning occurs when students are challenged intensively, leading to upper-level intellectual and ethical development. For instance, when masters students learn to deal with uncertainty and relativism, they experience a significant positive emotional engagement in the learning activities (Lunevich, 2021; Vygotsky, 2016). Teachers try to establish and maintain supportive relationships and cooperative, collaborative learning goals, and minimize the pressures that dispose students towards performance goals or work avoidance goals (Lunevich, 2021). When these conditions are created in a classroom or a virtual classroom, students can focus their energies on learning without becoming distracted by fear of embarrassment or failure, or resentment of tasks they view as pointless or inappropriate.

The learning environment of transformative learning can be characterized by teachers assisting students to achieve a sense of flow; goals are clear and compatible, feedback is immediate, and challenges are easy, stretching students' limits (Lunevich, 2021; Sternberg, 2004). The whole learning environment is set up to expect students to succeed. High-order thinking, engaging, challenging learning processes occur as sharing love to teach and learn, sharing enthusiasm, and passion for social interactions, exchanging ideas and creating space for all to contribute. The next step in this process is the analysis of information by learners, problem-solving activities, analysing real events, and receiving feedbacks from the teacher. In this process, students engage in motivation and active learning with immediate feedback from the teacher. Transformative learning is deep and through constant change of learning activities, analysing information, and offering students new challenges.

For transformative learning to occur, the following conditions need to be achieved: (1) the course learning content should be well developed and suited to the course objectives; (2) teaching approach should align with assessments tasks and course objectives; and (3) level of student engagement should be appropriate to the course objectives. Further, trust between teacher and learners is the foundation of transformative learning.

Sternberg (2008, 2010), who wrote a wealth of literature on patterns of giftedness, pointed out that giftedness comes in many forms (Sternberg et al., 2008). Theorists have attempted to classify types of giftedness using various classifications (Sternberg, 2010). Sternberg described seven patterns of giftedness and specifically mentioned the types of environments in which gifted individuals make contributions. Renzulli (1998), cited in Sternberg (2000), distinguished between schoolhouse gifted individuals and creative-productive gifted individuals (Sternberg, 2000, 2006). According to them, schoolhouse gifted individuals are likely to show their gifts in academic environments, but many do not show gifts outside such environments. This is particularly important to curriculum design and education, as we want our students to succeed once they have completed the school or university degree. Conversely, according to Sternberg (2000), creative-productive gifted individuals are more likely to show their gifts outside academic environments and make more enduring contributions (Sternberg, 1997, 2008; Sternberg & Lubart, 1991; Sternberg et al., 1997).

Gardner (1999) suggested another path in proposing types of intelligence that can be characterized as gifted individuals (Gardner, 1983, 1999, 2000; Sternberg, 2001, 2019). He analysed the intelligence of 20th-century creative leaders and used the multiple intelligences to suggest that each gifted individual excelled in a different intelligence. For example, Mahatma Gandhi excelled in intrapersonal intelligence. T. S. Elliot in verbal intelligence, Albert Einstein in logical-mathematical intelligence, and specific intelligence could be extended to Leonardo da Vinci (Gardner, 1991)

Sternberg defined that giftedness is based not on types of intelligences but on patterns of intelligence (Sternberg, 2008). According to his triarchic theory, individuals have patterns of strengths and weaknesses, which can be loosely used for classification (R. Sternberg, 1985). A person may show certain patterns, but they may be at the intersection between patterns, and their patterns may change over time.

Gifted individuals whose contributions have been historically enduring have generally demonstrated a minimum of three common attributes that can be revealed by analysing their contributions. As noted above, Sternberg (1998, 1997) identifies these three attributes as analytical skills, the ability to analyse and evaluate one's own ideas and those of others; creativity skills, the ability to generate one or more major ideas that are novel and of high quality; practicality skills, the ability to convince people of the value of ideas and render the ideas practical (Sternberg, 1997, 2006). The kinds of tasks people face in everyday life often require all these abilities. Sternberg (2000) continued that different combinations of these skills lead to different patterns of giftedness, which he proposed in *Patterns of Giftedness: A Triarchic Analysis* (Sternberg, 2000, 1985; Sternberg & Lubart, 1996), (Rozesahegyi, 2019).

REFERENCES

Emmons, R. (2000). "Is Spirituality An Intelligence?". *The International Journal for the Psychology of Religion, 3*, 27–34.
Gardner, H. (1983). *Frames of Mind: The Theory of Multiple Intelligences.* New York: Basic.
Gardner, H. (1991). *Multiple Intelligences: The Theory in Practice.* New York : Basic.
Gardner, H. (1999). *Intelligence Reframed. Multiple Intelligences for the 21st Century.* New York: Basic Books.
Gardner, H. (2000). A Case Against Spiritual Intelligence. *The International Journal for the Psychology of Religion, 10*(1), 27–34.
Lunevich, L. (2021). Creativity in Teaching and Teaching for Creativity in Engineering and Science in Higher Education—Revisiting Vygotsky's Psychology of Art. *Creative Education*, 12, 1445–1457. doi: 10.4236/ce.2021.127110.
Rozesahegyi, T. (2019). Observations. In M. Lambert (Ed.), *Practical Research Methods in Education: An Early Researcher's Critical Guide* (pp. 23–34). London: Taylor & Francis Group.
Sternberg, R. (1985). *Beyond IQ: A Triarchic Theory of Human Intelligence.* New York: Cambridge University Press.
Sternberg, R. (1991). An Investment Theory of Creativity and Its Development. *Human Development, 34*, 1–34.
Sternberg, R. (1997). Creativity as Investment. *California Management Review, 40*, 8–21.
Sternberg, R. (1998). Human Abilities. *Annual Review of Psychology, 49*, 479–502.
Sternberg, R. (2000, June). Patterns of Giftedness: A Triarchic Analysis. *Roeper Review, 22*, 231.
Sternberg, R. (2001). What Is the Common Tread of Creativity? Its Dialectical Relation to Intelligence and Wisdom. *American Psychologist, 56*(4), 360–362.
Sternberg, R. (2004). Teaching College Students that Creativity Is a Decision. *Guidance & Counseling, 19*(4), 196–200.
Sternberg, R. (2006). The Nature of Creativity. *Creativity Research Journal, 1*, 394–414.
Sternberg, R. (2010). Teaching for Creativity. In R. Beghetto, & J. Kaufman (Eds.), *Nurturing Creativity in the Classroom* (pp. 394–414). New York: Cambridge University Press.

Sternberg, R., Kaufman, J., & Grigorenko, E. (2008). *Applied Intelligence*. New York: Cambridge University Press.
Sternberg, R., & Lubart, T. (1996). Investing in Creativity. *American Psychologist*, *51*(7), 677–688.
Sternberg, R., O'Hara, L., & Lubart, T. (1997). Creativity as Investment. *California Management Review*, *40*(1), 8–21.
Vygotsky, L. S. (2016). *The Psychology of Art* (E. T. Technology, Trans.). THE M.I.T PRESS, Cambridge, Massachusetts, and London, England.

Critical Digital Pedagogy – Innovative Model

3

Lucy Lunevich

3.1 INTRODUCTION

Critical digital pedagogy provides frameworks for the multitude of decisions teachers make to assess their own teaching and the digital literacy of their students. Like any kind of innovation, innovation in pedagogy takes new ideas and practices and brings them together in new ways to solve problems that presently do not have adequate solutions. Developing new pedagogical models, however, involves both the identification of problems and the suggestion of novel practices. The full power of pedagogy and pedagogical innovations can only be evaluated by considering all the things the pedagogies are trying to achieve. The present research evaluates various themes of critical pedagogy, suggests an innovative model, and focuses on intentions when evaluating pedagogies rather than assuming they all have the same purpose.

Theories of education consider the interactions between students and teachers, between the state and the educational system, and between universities and the economic, technological, and social development of society. Therefore, critical digital pedagogy must look beyond the internal relationships of the teacher and students, as well as their social environment. It must also be open to innovations, interferences, and dynamic changes. This would thereby form subject competencies and design the formation of social, communicative, and life competencies (K. Smith, 2013; M. Smith, 2021; Volov, 2007). Teaching and learning processes have several components: purpose and objectives, content, methods, teaching tools, learning forms, and results. To achieve teaching and learning objectives, all areas of critical pedagogy must be considered in the context of new environments and social and technological changes (Lunevich, 2021; Myamesheva, 2015). In this research work, a new concept of critical digital pedagogy is suggested based on research observations, personal communications, and recent conference presentations. Recent developments in pedagogy have been consolidated as an innovative model for higher education, as shown in Figure 3.1.

We live in an age of information: communication technologies and data economy are the primary tools of production that contribute to enhancing the social positions of students (R. Rogers, 2013; D. Smith, 2003). With online and blended learning and teaching practices becoming more common, the ratio of students per teacher can reach into hundreds (K. Smith, 2013; M. Smith, 2021). Critical digital pedagogy requires the development of innovations in education and pedagogy, and a new approach to teaching and learning (Lunevich, 2021; R. Rogers, 2013; J. Rosen, 2015). Although technical know-how is one aspect of digital literacy, curriculum documents tend to overlook social, cultural, and ethical issues related to learning with

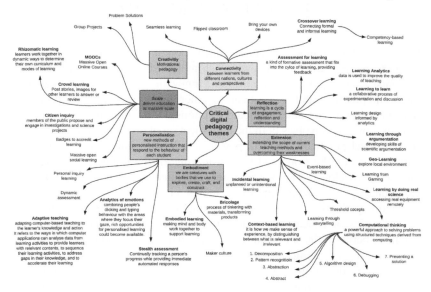

FIGURE 3.1 Critical digital pedagogy themes: the Lunevich model.

technology (Hadziristic, 2017; Hobbs, 2010). However, they do examine local and global responsibilities and professional practice (Vasquez, 2013; Volov, 2007). According to Hadziristic (2017) and K. Smith (2013), the main challenges educators face are insufficient opportunities for professional development to build digital pedagogical competencies and update curricula to create effective pedagogies (Hadziristic, 2017; K. Smith, 2013). The professional development of year-12 teachers in Canada has focused on developing digital skills and competencies and has consequently neglected to engage with the cultural and social contexts of technology use and integration (Haight, 2014; Howe, 2009). The lack of analyses of knowledge production and power relations in technology often results in educational practices that inadvertently repeat historical patterns of injustice and inequality (Haight, 2014; Kukushkina, 2002; MacNeill, 2003). Critical pedagogy should stress the mastery of not only technical skills but also critical thought (Haight, 2014). According to Hobbs and others, digital competencies must move from merely acquiring functional skills to solving community, regional, or national issues (Helsper, 2010; Hobbs, 2010; A. K. Mynbayeva, 2007; G. Mynbayeva, 2016).

Baround and Dharamshi (2020) have pointed out that digital technologies should be considered not only as tools for meaning-making but also as digital texts and platforms for literacy practices (Balasanyan, 2019; Baroud, 2020). This shift would encourage teachers to consider integrating critical digital literacies into curricula rather than treating them as an optional topic to be addressed. It is believed that such a paradigm shift would encourage educators to consider thinking critically, ethically, and responsibly, and design literacy-learning opportunities that respond to institutional, social, and economic contexts and requirements (Balasanyan, 2019; Hobbs, 2010; Howe, 2009; Lunevich, 2021).

3.2 RE-EMERGING PEDAGOGY

Smith and Smith assert that pedagogy is wrongly seen as the "art and science of teaching" (K. Smith, 2013; M. Smith, 2012). Pedagogy also involves joining with others to bring relationships to life (animation), being concerned about one's own and other's needs and well-being, taking practical steps to help (caring), and encouraging reflection, commitment, and change (education) (Kant, 1900; Kukushkina, 2002). Smith (2021) and PLATÃO (1988) added that education is a deliberate process of drawing out learning (Latin: *educere*) and of encouraging and giving time for discovery (PLATÃO, 1988; M. Smith, 2021). It is an intentional act, a process of inviting truth and possibility, and is "based on certain values and commitments such as respect for others and for truth" (M. Smith, 2021). According to Smith (2012), Smith (2003), Cope (2021), and Freire (2005), pedagogy has many different facets, including pedagogy of the oppressed, critical pedagogy, digital pedagogy, and social pedagogy (Lunevich, 2021).

Four main competence areas have also been identified: pedagogical, technological, collaborative, and creative (Kukushkina, 2002; Lunevich, 2021). According to Smith (2021), this "refers to competencies involved in making pedagogical choices throughout the process of teaching and learning" in a game-based context (K. Smith, 2013). Kukushkina (2002) also states that "pedagogy involves conveying knowledge and skills in ways that students can understand, remember", and apply, regardless of which learning style the learner has (Kukushkina, 2002). Twenty-first-century pedagogy aims to develop the skills and knowledge students need to succeed in work, life, and citizenship. Twenty-first-century skills can be applied in all subject areas and all educational, career, and civic settings throughout a student's life (A. S. Mynbayeva, 2012, 2018; G. Mynbayeva, 2016).

Lunevich (2021) pointed out that pedagogy is about connecting a learner's mind and heart (Lunevich, 2021). According to MacNeill, it is also a matter of grace and wholeness, wherein students can engage fully with their innate gifts and strengths (MacNeill, 2003). As Pestalozzi and others affirmed, education is rooted in human nature; it is a matter of "head, hand, and heart" (Bruhlmeier, 2010; Colligan, 2020; Pestalozzi, 2010). In life, each person develops their own sense of identity, meaning, and purpose "through connections to the community, to the natural world, and to spiritual values such as compassion and peace" (Kant, 1900; Kukushkina, 2002). Pedagogy is also conceptualized as a process of "accompanying learners; caring for and about them; and bringing learning into life" (Bordovskaya, 2000; Kupsevich, 1986). Plato and others made a distinction between education (Latin: *educatio*) and teaching (Latin: *instructio*). "Education means shaping the development of character with a view to the improvement of man" (PLATÃO, 1980a, 1988) while "[t]eaching represents the world, conveys fresh knowledge, develops existing aptitudes and imparts useful skills…" (Plato, 1925; Podlasy, 1996). Teaching is concerned not just with "knowing about" things but also with changing oneself and the world one lives in (Lunevich, 2021; Myamesheva, 2015; R. Rogers, 2013).

Within ancient Greek society, there was a substantial distinction between the activities of pedagogues (paidagögus) and subject teachers (didáskalos) (K. Smith, 2013). Plato talks about pedagogues as "men who by age and experience are qualified to serve as both leaders (hëgemonas) and custodians (paidagögous [sic])" of children (PLATÃO, 1980b). Their roles varied, but two elements were common between them. "The first was to be an accompanist or companion – carrying books and bags and ensuring their wards were safe" (PLATÃO, 1980a). The second and more fundamental task was to help boys learn what it was to be men. They accomplished this by a combination of example, conversation, and discipline. Pedagogues were moral guides who were to be obeyed (Bernstein, 1990; Kharbach, 2021). Employing a pedagogue was a custom that went far beyond Greek society: well-to-do Romans and some Jews also placed their children in the care and oversight of trusted servants (M. Smith, 2021). The relation of the pedagogue to the child is a fascinating one. It brings new meaning to Friere's (2005) notion of the "pedagogy of the oppressed" – the education of the privileged by the oppressed (Freire, 2005). It was a matter that, according to Plato, did not go unnoticed by Socrates (Plato, 1925). In a conversation between Socrates and a young boy, Lysis, Socrates asked, "Someone controls you?" Lysis replied, "Yes, he is my tutor [or pedagogue] here". "Is he a slave?" Socrates queried. "Why, certainly; he belongs to us", responded Lysis, to which Socrates mused, "What a strange thing, I exclaimed; a free person controlled by a slave!" (Plato, 1925, quoted by Smith, 2006).

The distinction between teachers and pedagogues, instruction and guidance, and education for work or life was a recurring feature of discussions around education for many centuries. It was still around when Immanuel Kant (1900) explored education (Kant, 1900). In *On Pedagogy* (Über Pädagogik), first published in 1803, he wrote that culture is positive, consisting of instruction and guidance (and thus forming part of education) (Kant, 1900). Guidance means directing the pupil in putting into practice what he has been taught. This illustrates the difference between a private teacher who merely instructs, and a tutor or governor who guides and directs his pupil. The former trains the student for a job only; the latter trains the student for life (Kant, 1900; M. Smith, 2021).

One of the important landmarks in pedagogical research was the publication of John Amos Comenius's book *The Great Didactic* [Didactica Magna] (first published in Czech in 1648, Latin in 1657, and English in 1896). For Comenius, the fundamental aims of education generated the basic principle of Didactica Magna, omnis, omnia, omnino – to teach everything to everybody thoroughly, in the best possible way (M. Smith, 2021). Comenius believed that every human being should strive for perfection in all that is fundamental for life and do this as thoroughly as possible (Lunevich, 2021; M. Smith, 2012). According to Colligan (2020), every person must strive to become (1) a rational being, (2) a person who can rule nature and their own self, and (3) a being mirroring the creator (Bordovskaya, 2000; Colligan, 2020). He developed sets of rules for teaching and laid out basic principles. His fundamental conclusions, according to Colligan (2020), remain valid: (1) teaching must be in accordance with the student's stage of development;

(2) all learning happens through the senses; (3) one should teach in order from the specific to the general, from what is easy to what is more difficult, from what is known to what is unknown; (4) teaching should not cover too many subjects or themes at the same time; and (5) teaching should proceed slowly and systematically. Nature makes no jumps (Colligan, 2020). Later in this paper, this concept is tested.

Following Kant and Comenius, another significant turning point in thinking about teaching came through the growing influence of one of Kant's successors as the Chair of Philosophy at the University of Königsberg, Johann Friedrich Herbart (1776–1841) (Freire, 2005). Like practical and theoretical educationists before him, Herbart distinguished between education (Latin: educatio) and teaching (Latin: instructio). "Education", according to Hebart, meant shaping the development of a student's character with a view to their personal improvement, while "teaching" represented the world, conveyed fresh knowledge, developed existing aptitudes, and imparted useful skills (Freire, 2005). Before Herbart, it was unusual to combine the concepts of "education" and "teaching". Consequently, questions pertaining to education and teaching were initially pursued independently. In his educational theory, Herbart took the bold step of subordinating the concept of teaching to that of education. As he saw it, external influences, such as the punishment or shaming of pupils, were not the most important instruments of education (Freire, 2005; Dewey, 1916). On the contrary, appropriate teaching was the only sure means of promoting education that was bound to prove successful.

In Aristotle's terms, pedagogy comprises a leading idea (eidos), what we are calling haltung or disposition (phronesis – a moral disposition to act truly and rightly), with dialogue and learning (interaction) and action (praxis – informed, committed action) (D. Smith, 2003). In the following summary, many of the elements that have been explored here are present (D. Smith, 2003; M. Smith, 2012). To this, we must add what Aristotle discusses as hexis – a readiness to sense and know. This is a state of being, or what L. Lunevich 2021) talks about as an "active condition". It allows us to take a step forward, both in terms of the processes discussed above and in what we might seek to do when working with learners and participants. Such qualities are conceptually at the core of the haltung and the processes of pedagogues and informal educators.

Within critical pedagogy, there is a strong emphasis upon being in touch with one's feelings, attending to one's intuition, and seeking evidence to confirm or question what one might be sensing. A further element is also present: a concern to not take things for granted or simply accept them at face value. Figure 3.1 shows several themes of the critical digital pedagogy and innovative model suggested.

The research and history of pedagogy from Plato to the modern day necessitate that critical pedagogy must consider all of these different themes of creativity simultaneously: connectivity, reflection, extension, embodiment, personalization, and scale. This approach helps create learning content and learning environments that consider all learning styles and learners' various social and cultural backgrounds.

3.3 INNOVATIVE MODEL FOR THE DIGITAL ERA

Research indicates that the critical features of a 21st-century education include personalized learning; equity, diversity, and inclusivity; learning through doing; the changed role of the teacher; community relationships; technology; and teacher professionalization (Pestalozzi, 2010; R. Rogers, 2013; K. Smith, 2013; M. Smith, 2021). According to Smith (2021), teacher "preparation and professional development should be reworked to incorporate training in teaching key competencies" (M. Smith, 2021). The 21st-century teacher must know how to provide technologically supported learning opportunities for students and understand how technology can support learning (A. S. Mynbayeva, 2018; M. Smith, 2021). Specifically, the five steps pertaining to this are preparation, presentation, association, generalization, and application (M. Smith, 2021). This "suggests that pedagogy relates to having assumptions as an educator and a specific set of abilities with a deliberate end goal in mind" (M. Smith, 2021; So, 2013; G. O. Soldatova, 2015).

However, critical digital pedagogy is concerned with engaging students in all learning styles, assuming that any individual has a minimum of two or three learning styles (M. Smith, 2021). These styles include seven types that must be considered when courses are offered online and in a blended mode (face-to-face and online) (Table 3.1).

Most learners have several learning styles, rather than simply one or two (Sitarov, 2008; G. R. Soldatova, 2016; Taubayeva ShT, 2001). Therefore, a challenge for critical digital pedagogy is to consider all learners' learning styles and design courses, programs, and learning environments and adopt engagement strategies that include all learners (Lunevich, 2021; A. S. Mynbayeva, 2012; M. Smith, 2021). Critical digital pedagogy is an emerging discipline within pedagogy that seeks to study and use contemporary digital technologies in teaching and learning, although this might not always consider all learning styles and learners' requirements to progress well. Critical digital pedagogy may be applied to online, hybrid, and face-to-face learning environments.

Critical digital pedagogical skills for teachers "thus include the capacity to plan, initiate, lead, and develop education and teaching with the departure

TABLE 3.1 Learning styles

LEARNING STYLE	PREFERRED LEARNING ENVIRONMENTS
Visual (spatial) learner: prefers learning by observing things	Face-to-face, online
Aural (auditory) learner: prefers to learn face-to-face in class	Face-to-face, classroom, workshops
Verbal (linguistic) learner: prefers a style that involves both the written and spoken word	Face-to-face, online
Physical (kinesthetic) learner: often referred to as "learning with the hands" or physical learning	Face-to-face, classroom, workshops
Logical-mathematical learning style: able to reason, solve problems, and learn using numbers, abstract visual information, and analysis of cause-and-effect relations	Face-to-face, online
Interpersonal learner: learns best through communication with others, whether verbal or nonverbal. Interpersonal learners love being around people and working in groups or teams and overall thrives through social interactions. They are often seen as social butterflies because they like spending much of their time with others. Interpersonal learners love sharing their knowledge with others but also love listening to their peers.	Face-to-face, classroom, workshops
Solitary learner: also known as an intrapersonal learner. Prefers learning on their own. They are self-motivated, enjoy working independently, and learn best when working alone. Solitary learners spend much time in self-reflection and enjoy working on themselves. They also favour a quiet environment in both their personal and academic lives.	Online, blended learning environment

point in both general and subject-specific knowledge of student learning" (English, 2015; Hadziristic, 2017; Hegenan & Olson, 2004). Critical digital pedagogical skills also include assessing students' digital skills and learning styles and then connecting the teaching to research on the subject of interest (A. S. Mynbayeva, 2012; R. Rogers, 2013). Critical pedagogy is the relation between learning techniques and culture and social environment need to be considered in teaching (MacNeill, 2003; Mynbayeva, 2007). It is determined based on an educator's beliefs about how learning takes place. Pedagogy requires meaningful classroom interactions between educators and learners (Lewin & Lundie, 2016; Lunevich, 2021; G. Mynbayeva, 2016). The goal is to help students build on prior learning and develop skills and attitudes regardless of whether they are in class or online. Critical digital pedagogy must be culturally relevant: it must focus on multiple aspects of student achievement and support students to uphold their cultural identities (Kukushkina, 2002; Kupsevich, 1986). Culturally relevant pedagogy also calls for students to develop critical perspectives that challenge societal inequalities (Lunevich, 2021; M. Smith, 2021). Furthermore, the five principles of pedagogy, i.e. motivation, exposition, direction of activity, criticism, and inviting imitation, will need to be considered in learning design (G. Mynbayeva, 2016; R. Rogers, 2013; G. R. Soldatova, 2016).

3.4 DIGITAL LEARNING

In the paper "Critical pedagogies in the neoliberal university: What happens when they go digital?", Smith and Jeffry (2013) discussed the "successful" online education performance (K. Smith, 2013). As a technology of reform, online education creates intense structural change within postsecondary institutions and reconstitutes what it means to teach critically and engage learners in multiple different ways (Lunevich, 2021; G. Mynbayeva, 2016; A. S. Mynbayeva, 2018; Mukazhanova, 2013). Although it can be argued that much of academic labour has always been hidden and unrecognized, the introduction of online teaching practices substantially increases invisible work by approximately 80%, according to our research and research conducted by others over the past two years. Online teaching entails an intensification of what is expected of educators within a space of work without boundaries (Drewey, 1916, Lunevich, 2021; K. Smith, 2013). Overwhelming reports from universities and academics describe online teaching as a constant struggle – night and day, seven days a week – to stay on top of responses to emails and monitor students' posts on course sites. Far more than a problem of poor boundaries, these work practices must be seen alongside rising pressures to be "always on the job" (K. Smith, 2013). As Hobbs (2010) found, online technology can monitor faculty availability, activities, and responsiveness to student demands (Hobbs, 2010). Some warn that these heightened surveillance measures could result in loss of autonomy and increased scrutiny, risking the creation of conditions leading to a rise in self-censorship and potential loss of academic freedom (K. Smith, 2013). Just as surveillance mechanisms transform the relationship between educators and their workplaces, the introduction of commercialized instructional products promises to radically alter how they engage with students. Online tracking of student learning should provide an instructor with the knowledge of how well they integrate unsettling course content such as theorizations of antiracism and the online effects of colonialism. The result is that students are treated not as learners but as users of software data collection technology (K. Smith, 2013). Smith and Jeffrey pointed out that if education is to be efficient, then it simply must be capable of being measured (K. Smith, 2013). However, critical digital pedagogies do not lend themselves readily to easy measurement in this way.

How could this spatial shift from critical pedagogies to critical digital pedagogies offer better values to learners and educators? What should an innovative model look like? To respond to these questions, we must revisit Plato and Kant. Plato said that "knowledge is the food of the soul" and "writing is the geometry of the soul"; however, "knowledge becomes evil if the aim be not virtuous". This is the ideal of education expressed by the concepts of truth, goodness, and beauty (PLATÃO, 1988; Roberts, 2013).

In the environmental approach, information and energy become important categories. During lessons, there is a dynamic exchange of information, knowledge, and energy between teacher and student. According to Mukazhanova, the value of "love" is understood as an energy exchange between people, for example, mother and child (Mukazhanova, 2013). Positive attitudes in study and occupation, such as the positive energy generated by the teacher, set a special positive atmosphere for learners (A. K. Mynbayeva, 2007; G. Mynbayeva, 2016; A. S. Mynbayeva, 2018). Plato stressed the importance of "platonic love" or "spiritual communication between teacher and student'" (Plato, 1925). Therefore, in critical digital pedagogy, positive motivating methods should be used in order to create an environment that is positive for learning. This environment becomes saturated with both information and positive energy (Bodrova, 2007; Lunevich, 2021). The teacher simultaneously teaches and learns from students (A. S. Mynbayeva, 2012; M. Smith, 2021). As the ratio of students to teachers reaches hundreds, there are tens of thousands of instances of digital technology being used for teaching, critical digital pedagogy calls for innovations in education, pedagogical innovations, and a new approach to teaching and learning. As A. S. Mynbayeva, 2012; pointed out, social and digital participation is a "new concept of the practice of informal, socially-digital mediated participation" (Sweeney, 2014). According to Soldatova's and Zotova's research, changes have occurred in the digital generation's memory, attention, and thinking (Snyder, 2011, A. S. Mynbayeva, 2012). The average length of time for which students can concentrate, compared to that of 10–15 years ago, decreased 10 times from 120 minutes to 12 minutes (Soldatova & Rasskazova, 2014, 2015, 2017). According to Mynbayeva et al. (2018), "a new phenomenon is clip thinking". It is based on the fragmentary processing of visual images, rather than "on logic and text associations" (Soldatova & Rasskazova, 2014,

2015, 2017). Digital technologies have changed our way of life and ways of communication, as well as our thinking, feeling, and connecting to others. They also provide channels to influence other people, social skills, and social behaviour (K. Smith, 2013; Sweeney, 2014).

There is something important in the dynamics of teaching and learning that needs to be carried forward from old critical pedagogy to critical digital pedagogy; this is the "spiritual communication between teacher and student" (Lunevich, 2021; PLATÃO, 1980b). This aspect creates new requirements for teachers and their professional development. Teachers must learn new information and new technologies quickly, and a creative approach to teaching and connecting the "mind and hearts" of students is an essential aspect of this new social environment (Lunevich, 2021; Mukazhanova, 2013; Podlasy, 1996; K. Smith, 2013). In this context, teachers are responsible for encouraging, facilitating, directing, and monitoring the progress of online interactions among students during and between classes. This is knowledge obtained through interaction with the network community. Such a process of obtaining knowledge is characteristic of an already prepared or adult person who can critically evaluate, analyse, choose, and construct knowledge.

In education, and particularly in higher education, learning has shifted from acquiring knowledge and skills to forming competencies and practical applications of that knowledge and value for money. As a result, critical digital pedagogy should integrate several methods of communication and types of instruction. It is obvious that critical digital pedagogy calls for the development of innovations in education or new pedagogical models. However, innovation is a phenomenon that carries the essence, methods, techniques, technologies, and content of the new (Lewin & Lundie, 2016). According to *Pedagogy of the Twenty-First Century: Innovative Teaching Methods*, "innovative teaching methods involve new ways of interacting between teacher and student and a certain innovation in practical activity in the process of mastering educational material" (Balasanyan, 2019; J. Rosen, 2015). The key to critical digital pedagogy is that the teacher simultaneously teaches and learns from the students.

3.5 CONCLUSION

The methodology of teaching is built on a personal-oriented or student-centred approach to learning. The approaches it synthesizes can be described as synergetic, systematic, competent, dialogical, activity-oriented, culturological, informative, technological, and environmental. UNESCO

(2010) has recommended the following teaching strategies: experimental learning, storytelling, values education, enquiry learning, and community problem-solving (UNESCO, 2010; Babansky, 1983). Creative and innovative teaching approaches are becoming increasingly necessary: the more exciting and diverse they are, the better. These teaching strategies also shape the experience of solving nonstandard problems, promote in-depth training, and aid in the steady assimilation of the technology of practical activity. A good teacher constantly improves their own pedagogical skills and selects and develops new methods and technologies of teaching. A good university constantly monitors the quality of its education and the value it delivers to learners and the economy. Universities currently tend to have their own business models which are supposed to deliver unique values to the community, connect with the external environment, and find various funding opportunities.

REFERENCES

Babansky, Y. (1983). *Pedagogy: Textbook for Students of Pedagogoc Universities*. Moskow: Prosveshenye.
Balasanyan, B. (2019). Lessons from the History of Pedagogical Methods for Culturally Responsive Teaching and Learning. In S. Balasanyan (Ed.), *Education Systems around the World* (pp. 9–23). London: IntechOpen.
Baroud, J. D. (2020). A Collaborative Self Study of Critical Digital Pedagogy in Teacher Education. *Studying Teacher Education 16*(2), 164–182.
Barry, W. A. (1986). *The Practice of Spiritual Direction*. New York: Harper and Collins.
Bernstein. (1990). *The Structuring of Pedagogical Discourse. Class, Codes and Control*. (Vol. 4). London.
Bodrova, E. L. D. (2007). *Tools of the Mind: "Vygotskian approach to early childhood education"*. (M. Hall, Ed.) New York: Columbus.
Bordovskaya, N. R. (2000). *Pedagogy: Textbook for Students of Pedagogic Universities*. Saint Petersburg: Publishing House Piter.
Bruhlmeier, A. (2010). *Head, Heart and Hand. Education in the Spirit of Pestalozzi*. Cambridge: Sophia Books.
Colligan, C. A. (2020). *Notes from the Field: Student Perspectives on Digital Pedagogy*. New York: Digital Studies.
Cope, B. K. (2021, June 10). *Ubiquitous Learning: An agenda for Educational Transformation*. Retrieved from ResearchGate: https://www.researchgate.net/publication/228347819_Ubiquitous_learning_An_agenda_for_educational_transformation/link/00b7d51b7a236ad703000000/download
Dewey, J. (1916). *Democracy and Education. An Introduction to the Philosophy of Education*. New York : Macmillan.
Dewey, J. (2008). Experience and Education. *Journal of the Educational Forum*, 241–252.

English, L. (2015). STEM: Challenges and Opportunities for Mathematics Education. *Proceedings of the 39th Meeting of the International Group for the Psychology of Mathematics Education. 1*, pp. 4–18. Brisbane: IGPME - The International Group for the Psychology of Mathematics Education.

Freire, P. (2005). *Pedagogy of the Oppressed*. New York: Continuum.

Hadziristic, T. (2017, 04 17). *The State of Digital Literacy in Canada: A Literature Review: Brookfield Institute for Innovation and Entrepreneurship*. Retrieved July 2, 2021, from https://brookfieldinstitute.ca/wp: https://brookfieldinstitute.ca/wp-content/uploads/BrookfieldInstitute_State-of-Digital-Literacy-in-Canada_Literature_WorkingPaper.pdf

Haight, M. Q.-H. (2014). Revisiting Digital Divide in Canada: The Impact of Demographic Factors on Access to the Internet, Level of Online Activity, and Social Networking Site Usage. *Information, Communication and Society, 17*(4), 503–519.

Hegenan, B., & Olson, M. (2004). *The Theory of Learning*. Saint Petersburg: Publishing House Piter.

Helsper, E. (2010). Digital Natives: Where Is the Evidence? *British Educational Research Journal, 36*, 503–520.

Hobbs, R. (2010). *Digital and Media Literacy. A Plan of Action*. The Aspen Institute. London: The Aspen Institute.

Howe, N. S. (2009). *Millennials Rising: The Next Great Generation*. Boston, MA: Vintage Books, Random House.

Kant, I. (1900, 10). *Kant on Education*. Retrieved 06.23, 2021, from http://files.libertyfund.org/files/356/0235_Bk.pdf: http://files.libertyfund.org

Kharbach, M. (2021, June 22). *Educational Technology and Mobile Learning*. Retrieved from Andragogy and Pedagogy: https://www.educatorstechnology.com/2013/05/awesome-chart-on-pedagogy-vs-andragogy.html

Kukushkina. (2002). *Pedagogical Technologies*. Postov-on-Don: Publishing House Mart.

Kupsevich, C. (1986). *Fundamentals of General Didactics*. Moskow: Vysshaya shkola.

Lewin, D., & Lundie, D. (2016). Philosophies of Digital Pedagogy. *Studies in Philosophy and Education, 35*, 235–240.

Lunevich, L. (2021). Creativity in Teaching and Teaching for Creativity in Engineering and Science in High Education – Revisiting Vygotsky's The Psychology of Art. *Research Journal of Creative Education, 12*, 1445–1457.

MacNeill, N. (2003). Pedagogical Leadership: Putting Professional Agency Back into Learning and Teaching. *Curriculum and Leadership Education, 3*, 84–98.

Mukazhanova, R. O. (2013). *Self-Congnition Teaching Methods for Schools*. Almaty: Bobek NSPWC.

Myamesheva, G. (2015). The Virtue in the Modern Smart World. *Pedagogical Science, 44*(1), 152–156.

Mynbayeva, A. S. (2007). *Innovation Methods of Teaching, or How to Teach to Attract Students. Text Book*. Almaty: DOIVA.

Mynbayeva, A. S. (2012). *The Art of Teaching: Concepts and Innovative Methods of Teaching*. Almaty: Publishing House Kazakh University.

Mynbayeva, A. S. (2018). Pedagogy of the Twenty-First Century: Innovative Teaching Methods. In A. S. Mynbaeva (Ed.), *New Pedagogical Challenges in the 21st Century – Contributions of Research in Education* (pp. 20–45). London: IntechOpen.

Mynbayeva, G. (2016). Informatization of Education in Kazakhstane: New Challenges and Further Development of Scientific Schools. *International Review of Management and Marketing, 6*, 259–264.

Pestalozzi, J. (2010). *Leonard and Gertrude.* London & Memphis, TN: General Books.

PLATÃO. (1980a). *Iāo. 2. ed. Tradução de Carlos Alberto Nunes.* Belém: Edufpa.

PLATÃO. (1980b). *Hípias maior. 2. ed. Tradução de Carlos Alberto Nunes.* Paris : Belém: Edufpa,.

PLATÃO. (1988). *República. 2. ed. Tradução de Carlos Alberto Nunes.* Paris: Belém: Edufpa,.

Plato. (1925). *Lysis 208 C, trans WRM Lamd.* Cambridge, MA: Harvard University Press.

Podlasy. (1996). *Pedagogy: Textbook for Students of Pedagogic Universities.* Moskow: Vlados.

Roger, R. (2013). *Digital Methods.* Cambridge, MA; London, UK: The MIT Press.

Roberts. (2013). *Herbet, Dewey on Discountinuoty of Education.* New York: Cambridge University Press.

Rosen J, S. M. (2015). Open Digital Pedagogy - Critical Pedagogy. *City University of New York (CUNY)*, 2–12.

Siemens. (2021, 06 22). *Connectivism: Learning Theory or Pastime for the Self-Amused?* Retrieved from https://www.semanticscholar.org/paper/Connectivism%3A-Learning-Theory-or-Pastime-of-the-Siemens/9b476f6d47e12f292e7bce21f10585f3ffcd7f16; https://www.semanticscholar.org/paper/Connectivism%3A-Learning-Theory-or-Pastime-of-the-Siemens/9b476f6d47e12f292e7bce21f10585f3ffcd7f16

Sitarov, V. (2008). *Didaktika: Textbook.* Moskow: Publishing House Academy.

Smith, D. Lovat, T., (2003). *Curriculum Action on Reflection.* Boston, MA: Cengage Learning.

Smith, K. & Jeffry, D. (2013). Critical Pedagogy in the Neoliberal University: What Happens When They Go Digital?. *The Canadian Geographer, 57*(3), 371–380.

Smith, M. (2012, Oct 08). *What Is Pedagogy?* Retrieved June 22, 2021, from infed.org: education, community-building and change: https://infed.org/mobi/what-is-pedagogy/

Smith, M. (2021, June 15). *infed.org: Education, Community-Building and Change.* Retrieved from infed.org: education, community-building and change: www.infed.org: education, community-building and change

Snyder, C., Lopez, S., & Pedrootti, J. (2011). *Positive Psychology: The Scientific and Practical Explorations of Human Strengths.* Thousand Oaks, CA: Sage Publications, Inc.

So, W. W.-M. (2013). Connecting Mathematics in Primary Science Inquiry Projects. *International Journal of Science and Mathematics Education, 11*, 385–406.

Soldatova, G. O. (2015). Attitude to Privacy and Protection of Personal Data: Safety of Russian Children and Adolescents. *National Psychological Journal* (3), 56–66.

Soldatova, G. R. (2016). "Digitial Divide" and Intergenerational Relations of Children and Parents. *Psikhologicheskiy zhurnal, 37*(5), 44–554.

Sweeney, E. (2014, June 10). *Digital Australia.* Retrieved from Digital Australia: https://digitalaustralia.ey.com/?utm_content=buffer8ab23&utm_medium=social&utm_source=twitter.com&utm_campaign=buffer

Taubayeva ShT, L. S. (2001). *Pedagogical Innovation as a Theory and Practice of Innovations in the Education System.* Almaty: Gylym.
UNESCO, (2010), Education Report, retrieved on Jan, 2022. https://www.unesco.org/archives/multimedia/document-504
Vasquez, M. F. (2013). *Technology and Critical Literacy in Early Childhood.* London: Routledge.
Volov, V. (2007). Innovative Principles of Education System. *Pedagogy, 7,* 108–114.

4 Creativity in Teaching and Teaching for Creativity – Engineering Students

Lucy Lunevich

4.1 INTRODUCTION

The chapter presents observational research conducted over four years which showed that creativity can be enhanced in all students using a variety of teaching methods and strategies. Strategies recommended to promote creativity include engaging students with activities that they enjoy and find intrinsically, encouraging students to verbally present and discuss their ideas with others, and providing opportunities for students to develop arguments based on evidence. Students should be encouraged to value empirical evidence and relevant knowledge and ideas, identify and address obstacles, learn about the lives and contributions of creative individuals throughout history, collaborate with others, take intellectual risks, and learn from mistakes in order to develop cognitive skills and self-efficacy. The research outlined here demonstrated that teaching critical analysis of art stimulated creativity in learners by facilitating emotional connections with the environment, encouraging a more holistic understanding of artworks and the creative processes behind them, and stimulating creative thinking skills and orientations that facilitated the development of intellectual potential.

4.2 CREATIVE PEDAGOGY

This chapter presents research conducted at the National Gallery of Victoria in Melbourne into methods that may facilitate students' greater engagement, learning, and creativity by exposing them to art installations, paintings, and modern and traditional art forms, and encouraging them to respond creatively to the stories behind the artworks. Vygotsky proposed that "creativity arises from any human activity that produces something new". Within his constructivist theory, the analysis of art has the potential to facilitate the development of higher cognitive abilities and provides an opportunity for collaborative learning when undertaken in group contexts. Vygotsky (2016) further proposed that many advances gained through human creativity and productivity have relied on collective creativity, in which small individual contributions combine to produce a greater outcome than individuals could have produced alone. Creativity of this kind should be encouraged among contemporary engineering and science students to support the problem-solving and critical thinking required to produce practical and theoretical solutions for complex modern problems (Treffinger et al., 2002).

Research indicates that modern workplaces demand critical thinking and creativity from employees, and these qualities are associated with higher productivity for employers and better work-life balance for employees (Bodrova & Leong, 2007; Skills, 2011). Employees who can process large amounts of disparate information and apply critical thinking to problem-solving, including scientists and engineers, represent a competitive advantage for businesses. In addition, these abilities are likely to extend the lifespan of businesses and increase their organizational capacity to recruit and retain more highly skilled employees. Critically, creative thinking is teachable, and the work of learning to think creatively is carried forward to be applied by learners across their later work (Karpov, 2014). That is, the work of learning creativity leads to the development of learners' creative abilities, which are then applied in the workplace (Vygostky, 2016). There is a consensus among scientists and scientist educators that scientific knowledge is the product of creative thinking (Marshall & Batten, 2003). Thus, the demand for greater creativity in education is due not only to its recognized pedagogical effectiveness but also to the demands of the global economy for flexibility and adaptation to deliver innovation and competitiveness (Marshall & Batten, 2003). However, empirical evidence has shown that there is a lack of appreciation of the importance of creative thinking among undergraduate engineering and science students (Wadaani, 2015). This paper synthesizes recent research that demonstrates how teaching creativity can facilitate human development and self-actualization for all students (Wadaani, 2015; Sternberg, 1985). The main goal of this research was to investigate the contribution of teaching for creativity to students' capabilities and intellectual potential (Lloyd, 2010; Wadaani, 2015) and provide evidence to inform higher education teaching practices in science and engineering.

According to Sternberg and Kaufman (1998) and Sternberg and Lubart (1991), students' potential abilities will not be developed if teaching and evaluation systems undervalue creative, analytical, and practical abilities. Teaching should include skills-based and informational content but should also include explicit instruction in strategies for the critical analysis and evaluation of received knowledge. As such, teaching should encourage and nurture creativity in diverse forms and facilitate the overall development of students to realize their full cognitive capacities. Sternberg (1985, 2006) explained that "teaching for creative as well as analytical and practical thinking combined enables students to capitalise on their strengths and to correct or to compensate for their weaknesses in order to be successfully intelligent individuals" (Torrance, 1995; Treffinger, 1988).

Similarly, Vygotskian theorists propose that children's development is a process that depends on adult mediation (Karpov, 2014; Vygotsky, 2016). As such, children will not develop to their full cognitive potential if teachers do not effectively scaffold their learning of skills and knowledge. The development of new interests and motivations in response to adult prompts leads students to engage with novel activities. Furthermore, the development of students' cognitive abilities – from their earliest developmental milestones to more advanced mental processes such as those needed for creativity – is proposed to rely on the acquisition, mastery, and internalization of psychological tools that are modelled by adults and more advanced peers, and whose learning is scaffolded by teachers. According to Vygotskian theory (Karpov, 2014), the cultural experiences shared with students through mediation or teaching practice are their "cultural heredity", and these experiences are not available outside of adult teachers. However, the teaching of science and engineering rarely includes teaching creativity in this way or teaching through creative arts. Therefore, this paper presents research conducted at the National Gallery of Victoria in Melbourne into methods that may facilitate greater engagement, as well as enhance learning and creativity, by exposing students to art installations, paintings, and modern and traditional art forms, and encouraging them to respond creatively to the stories behind the artworks.

4.3 ART IN TEACHING

To better understand the importance of creativity and creative arts for every individual, one must consider the significance art would acquire if this interpretation is correct. What is the relationship between aesthetic responses and other human behaviours? From a Vygotskian perspective, the function of art is to "infect" us with the emotions experienced by others, including through the stories behind the art (Karpov, 2014). Tolstoy proposed that "the activity of art is based on the capacity of people to infect others with their own emotions and to be infected by the emotion of others". Thus, observing art can lead us to discover, through imagination, a broader range of human experiences and knowledge than we could otherwise experience as individuals.

What connects teaching creative arts and teaching for creativity? They both provide learners with an emotional education. Following Tolstoy, teaching creativity involves not only imparting specific content knowledge and rational analysis but also scaffolding students' emotional and creative responses (Karpov, 2014; Vygotsky, 2016) suggested that creativity is at the heart of effective teaching itself and is vital to teaching in every subject area. They also proposed that learners need rich experiences to develop their creative skills and must be supported to use innovative approaches to enable this process of development. Therefore, teachers need to continuously reflect and incorporate learnings from their work into their teaching, as well as use the experiences of other teachers to develop fresh approaches to education that will inspire learners. Teachers need to move from flexibility in applying conventional methods to genuine creativity. This requires teachers to move out of their comfort zone of received knowledge into unknown territory to try new strategies and approaches (Davis, 2004; Lloyd, 2010).

There is a vast pedagogical literature on creativity that is relevant for university lecturers to apply in their teaching practice across subject areas (Woods, 1993, 1995). The role of creativity in teaching problem-solving strategies across a range of higher education contexts has been studied by several scholars, building on the work of Marshall and Batten (2003). For example, Woods (1990, 1993) argued that creativity is associated with the generation of unexpected responses through novel connections and associations between existing pieces of information and knowledge. This work is relevant for teaching science and engineering students to become effective problem-solvers in the modern workplace.

4.4 TEACHING FOR CREATIVITY

Creativity has been described as a multifaceted phenomenon that can be developed for all students in different areas (Wadaani, 2015, p. 6). Davis (2004) stated that "creativity can be expressed in a nearly infinite number of ways in human behaviour and has its origins in several components of individual and social experience". Rozesahegyi (2019) and Denscombe (2007) argued that creativity is an important component of problem-solving, healthy social and emotional well-being, and scholastic and workplace success. Therefore, teachers should not only teach creatively but also teach for creativity, thereby motivating students to think effectively and become creative learners who can make informed and considered decisions and choices in novel situations (Brinkman, 2010; Lloyd, 2010; Sternberg, 2006; Wadaani, 2015). As stated by Tanggaard (2013), "societies that do not make every effort to assure that the potential talents of young people are utilised are losing their most valuable natural resource; human capital" (Wadaani, 2015). Thus, teaching creativity is relevant across all subject areas, all disciplines, and for all students.

Research has also indicated that digital businesses in particular demand creative individuals who can solve their clients' problems and provide increased value for employers (Brinkman, 2010; Carroll, 1993). In business contexts, value is added by workers who can solve problems efficiently and demonstrate immediate commercial benefits for clients and employers through these solutions (Carroll, 1993). However, there is empirical evidence that many engineering and science students do not appreciate the relevance of creative thinking to their disciplines. Furthermore, teaching in science and engineering subjects can be dry and provide limited encouragement for students to analyse and apply content knowledge creatively.

Sternberg (1997) argued that creativity is not a learning objective to be achieved and measured, but an approach to thinking that all individuals should continue developing to the highest level possible. Teaching for creativity is not a teaching method, but a teaching philosophy designed to facilitate the continued development of students (Wadaani, 2015). Through this approach to teaching, multiple teaching strategies and methods can be modified or generated to lead students to develop their creativity in different contexts and at different levels. She and other scholars contend that teachers should teach for creativity as a complex capacity that can change and develop with infinite potential (Jeffrey & Craft, 2004; Wadaani, 2015). Most teaching methods can be adapted to develop student creativity and provide positive experiences for students within a philosophical framework of teaching for creativity.

When teachers understand their roles in constructivist terms as a facilitator of optimal human development and conceptualize creativity as "the hub of real achievement", they can continue to learn to teach for creativity by engaging in professional development (Wadaani, 2015). Teachers can practice novel teaching styles that complement the teaching methods they already use to create an environment that facilitates creativity. Teachers may also use teaching techniques designed to enhance creativity in general, or to foster and scaffold the development of certain aspects of creativity such as collective creativity (Falk & Szech, 2013; Wadaani, 2015). Here, teachers might benefit from taking risks and trying new strategies to evaluate what works.

4.5 CRITICAL THINKING AND PROBLEM-SOLVING SKILLS

Critical thinking has been described as the ability to engage in reflective and independent thinking (Skills, 2011) and has been the subject of debate and theorizing from the time of Plato and Socrates into the contemporary era. For example, critical thinking is often discussed in relation to students' ability to evaluate information found online (Karpov, 2014). Critical thinking requires students to use their ability to reason. Karpov (2014) defined critical thinking as the intellectually disciplined process of actively and skilfully conceptualizing, applying, analysing, synthesizing, and evaluating information gathered from, or generated by, observation, experience, reflection, reasoning, or communication, grounded in reality. Thus, critical thinking requires students to become active learners rather than passive recipients of information. In practice, this entails actively engaging with class conversations, being motivated to contribute and ask questions, and pursuing additional learning outside of classes. Critical thinkers rigorously question ideas and assumptions rather than accepting them at face value. They seek to evaluate whether ideas, arguments, and findings are biased or incomplete, and are open to challenges to accepted thinking. Critical thinkers will identify, analyse, and solve problems systematically rather than by intuition or instinct, or by deferring to authority.

Students who master critical thinking will be equipped to solve emerging business problems once they are employed.

4.6 OBSERVATION IN EDUCATION RESEARCH

As a method for education research, observation offers a way to analyse human behaviour and thinking in real-world contexts. Gathering useful observational data over time requires considerable methodological and ethical deliberation. For educators, carefully planned and rigorous observational protocols are recommended to usefully inform their professional practice. According to Rozesahegyi (2019) and Cohen (2018), analysing classroom observations is invaluable for comprehending and evaluating learners' knowledge, skills, curiosity, and cognitive development. In addition, the observation of teaching practitioners by other practitioners as a part of training and mentoring is invaluable for improving teachers' professional practice.

When education researchers conduct an observational research study, the observations must be planned and should follow specific protocols, and will be qualitatively and quantitatively different to everyday observing and perceiving in social contexts (Rozesahegyi, 2019). Cohen (2018) and Rozesahegyi (2019) pointed out that observational research involves "more than just looking, and requires systematic observation of the number and type of relevant events, behaviours, setting, artefacts, and routines". Observational research requires training and practice (Stanley, 1991).

In education research, observation may include the physical environment and organizational context, classroom organization, individual or group activities, allocation of teaching assistances in lesson time, the nature of interactions between participants, and the complexity of problems students are asked to resolve. Observations may also be made of pedagogical characteristics of the research setting such as the teaching strategies and resources adopted (Rozesahegyi, 2019). Observational research was conducted with child learners in primary schools from 1990 that initially focused on the creativity of the teacher and the nature of creative teaching (Weisberg, 1999). More recent research has focused on the effects of creative teaching on learners to evaluate its effectiveness by comparing the creativity they bring to the learning context with the creativity they were encouraged to develop through creative teaching activities (Sternberg, 1991, 2000, 2004, 2006, 2010 Weisberg, 1999; Woods, 1990, 1993, 1995). This paper takes its approach from these long-term ethnographic studies carried out in creative classes to investigate the relationship between teaching art and creativity, and teaching *for* creativity in education.

In this research project, the research questions were "how does positive emotional engagement of students influence the quality of learning?" and "does this pedagogical approach facilitate creativity in students with limited experience of art galleries?" Once an observational approach was chosen as an appropriate method of gathering data, research planning addressed practical issues including decisions about the role of the observer, how the processes of observation might influence the behaviour of those being observed, how the observation data will be recorded, analysed, and interpreted, and whether the observation method is ethical and morally sound. A further question was included for determining the specific focus of observations.

Structured or "systematic" observational methods were used in this research (Rozesahegyi, 2019). Observational data were recorded in defined categories that were designed to capture relevant behaviours, events, and activities. The researcher aimed to log the number and types of behaviours, events, and activities that allowed comparisons to be made between participants, as well as frequencies, patterns, and trends to be noted and evaluated (Cohen, 2018; Rozesahegyi, 2019). Observations included the gallery setting, student behaviour, and students' verbal responses to the artworks.

4.7 INVOLVEMENT OF STUDENTS IN EDUCATION ENQUIRY

Involvement of students in education enquiry has been advocated for many years. Rozesahegyi (2019) suggested that significant knowledge gains result when children's active participation in the research process is deliberately solicited and when their perspectives, views, and feelings are accepted as genuine, valid evidence. Rozesahegyi (2019) has argued that students have valuable perspectives to share about many issues, and that students' participation can make essential contributions to longer-lasting, longitudinal research. It would be beneficial to evaluate how the age, experience, and/or culture of students may bring greater variation.

Students were involved in this research in three ways. Over 200 students completed a four-page questionnaire which asked about their attitudes towards difficult versus easy academic tasks, and how creativity could help them to solve problems. In addition, students took part in small-group interviews, and individual questions were asked to all students in the gallery across various installations to gain a better understanding of their interpretation of the art form such as "What feelings does this stimulate?" It has been observed that the less experienced a student was with art, the more creative and innovative ideas they generated (Treffinger et al., 2013).

Features of the students' relationship to the investigation became particularly clear in the interviews. These interviews did not follow a consistent, formal, question-and-answer structure but were organized more around themes, exploring experiences and beliefs in a discussion style. For the most part, the participating students were self-assured enough to take control over the interview agenda, expressing opinions, spontaneously recollecting memorable occurrences, and even setting out new directions for dialogue. Although nothing inappropriate was said, the fact that the students were away from the lecturers may also have helped to promote active and honest discussion.

The fact that the discussions were carried out in groups of three, four, or five students, not individually, seemed to add to the value of the methodology. Fellow students, more than lecturers, stimulated judgments and reflections in each other.

4.8 RESEARCH METHODOLOGY

This observational study investigated teaching for creativity in lecture tours conducted with Masters students from RMIT University, College of Science, Health and Engineering, at the National Gallery of Victoria (NGV) in Melbourne over four years from 2018 to 2021. The program aimed to increase the engagement of international students. Participating students were in the age group of 23–53 years, and approximately 30% female and 70% male. Students in this program included international students from China, India, Indonesia, Malaysia, Norway, Turkey, Saudi Arabia, and Middle Eastern countries. Lecture tours were designed in collaboration with an NGV Curator.

Observational data were collected by the author, who delivered the program with NGV curators. Observational data collection focused on student behaviour and students' responses to questions during lecture-tour activities that were designed to engage students with the chosen artworks and their backstories. Observational data were collected of learners' unprompted expression of ideas and their participation in group activities. Students' emotional engagement with gallery installations was also noted. Individual students and student groups were asked to produce creative descriptions of unfamiliar artworks and imagine the story behind the artwork's creation. A main objective of the lecture tours was to empower learners to take ownership of their learning and provide educational activities in a setting that would stimulate emotional expression, individual connection to the artwork, creative evaluation of the artwork, critical thinking, and imaginative thought.

4.9 ETHICS OF OBSERVATION

This brings us to a reflection on ethics. Ethics can involve the basic idea that "research should avoid causing harm, distress, anxiety, pain any other negative feeling to participants", or even that researchers should aspire "to conduct research that benefits participants in positive ways". Indeed, the unusualness, the specialness, perhaps the "group"-ness of the task, especially the interviews, seemed to be a stimulus and reward enough for the students in this research. The lending of self-esteem, as well as the educational benefits which might ultimately result from the study and their participation in it, was well-balanced with the advantages of their involvement to the researcher himself. The building of "rapport" is often put forward as best practice in any kind of research in which others are involved, most credibly perhaps by Baker (2006) in relation to social work and police investigations.

Observation in the gallery raised its own ethical issues, particularly in relation to the researcher's relationship to the context being observed. Baker (2006) reminder is that the researcher should not challenge the accepted customs and value systems of the context of "social ecology", not just for ethical reasons, but also so that data will reflect the real nature of the observed setting. Instead, the researcher should try to merge with these systems. Patton (2015) has similarly advised that the researcher not disturb the relationship between students and lecturers.

4.10 RESULTS

Fifty percent of participating students had never previously been in an art gallery, 40% reported limited exposure to art history, and less than 10% of participating students had occasionally visited the gallery. Observation of students' ability to express their creative ideas in response to art installations in the NGV indicated that students engaged positively with the artworks, and their stories and were highly engaged when asked to express creative responses to complex questions posed by the lecturer and curator. For example, students engaged in animated discussion about their emotional responses to specific artworks. Many students offered multiple responses to the questions asked by the lecturer and tried to approach problems from different perspectives than those the lecturer had presented or considered.

All participating students demonstrated an ability to generate innovative responses, took ownership of the inquiry-based activities, offered varied solutions to questions posed by the lecturer and curator, and thought outside the norm. Students also demonstrated an ability to generate and evaluate ideas quickly during the limited time available. In these activities, students demonstrated the development of a more holistic understanding of the artworks, and the creative processes behind them, through true creative thought. Consistent with Vygotsky's constructivist theory, collective creativity was also observed in group activities when individuals pooled and discussed their creative responses to develop more advanced ideas and understandings. These findings are consistent with observations from the long-term ethnographic studies of child learners in creative classes discussed above.

In the NGV lecture-tour classes, students showed appreciation for the creative environment. This was demonstrated through their expressions of positive emotion, and through their enthusiasm in discovery and experimentation when perceiving and critically analysing the artwork, technology, and design. The NGV installations provided a focus for intensive problem-solving activity and critical thought. Students worked through frustrations experienced in the challenging process of developing novel and creative solutions to questions. They also expressed satisfaction in producing solutions and making conclusions based on critical analysis. Discussing the artworks with student peers provided an opportunity for students to express emotional responses and share those of their peers, facilitating collective creativity.

In addition, students were personally engaged with the artworks beyond the minimal requirements of lecture-tour participation. According to Patton (2015), ensuring the relevance of the pedagogy to learners, learning process is organised in ways that it will encourage their ownership of their engagement with learning, as "the learning will be directly related to their intrinsic interests". The lecturers scaffolded students' learning by suggesting interpretive strategies for analysing art that effectively engaged the students. Students then acted creatively to apply these strategies appropriately. The optimal pedagogical relevance of the NGV lecture tours to learners was evident in the way learners took ownership of their experiences in the NGV.

These findings support the idea that lecturers should aim to creatively develop materials and approaches that encourage students' interests and motivate their learning. Jeffrey and Craft (2004) suggested that teachers need to make teaching and learning relevant and encourage ownership of learning by passing back control of the learning activities to the learner (Amabile, 1982, 1996, 2001; Baer, 2010; Batey & Furnham, 2006; Jeffrey & Craft, 2004), and encourage innovation and critical thinking through emotional connections to the content. In the lecture tour, this was achieved through providing the opportunity for students to engage with artworks and their stories, to explore their preferences between the artworks on display, and to pursue critical thinking based on their perceptions of the artworks. Having control over learning provides learners with the opportunity to solve problems independently using their preferred strategies, and to innovate and express personal and developing ideas and feelings spontaneously in a supportive, collective context (Torrance, 1962, 1993, 1995; Treffinger, 1988, 1991).

One of the major characteristics of creativity is considering novel possibilities and exchanging ideas with others. Therefore, teaching for creativity should encourage learners to take control to facilitate innovation. In this research, learners were stimulated to think creatively by providing opportunities for emotional connection to artwork in an environment that supported expressions of individual and collective creative thinking in response to the subject material. Lecturers were available to listen and reinforce learners' creativity and created a supportive environment for learners to think creatively, solve problems, analyse and question the art form, and communicate about the art and its stories with other learners. Importantly, learners were encouraged to discuss ideas and stimulate creativity in others. Students were prompted to evaluate their creative ideas in discussion with others and develop further ideas collectively. These activities included teachers and students as co-participants, supporting research by Amabile (1982, 1983, 1996) which suggested that being encouraged to pose questions, [and] identify problems and issues together, with the opportunity to debate and discuss their "thinking", [takes] the learner into the heart of both the teaching and learning process as a co-participant (Cohen, Manion & Morrison, 2018). In this way, the lecture tours effectively encouraged students to take ownership of their learning and to practice creative thinking skills collectively (Amabile et al., 1996).

4.11 CONCLUSION

The observations reported in this research highlight that, regardless of age, cultural background, or gender, all students engaged emotionally with the artworks and their stories. The positive impact of teaching for creativity was demonstrated through learners expressing pleasure in experiencing the spark of creativity, exploring creativity by evaluating their own and others' ideas, and collectively developing new responses. The research outlined here shows that creative teaching of the arts stimulates creativity in learners by facilitating emotional connections with the artworks and the stories of their creation that engage learners. Importantly, these activities also stimulated intrinsically motivated engagement in activities that provided students with opportunities to develop creative thinking skills that could then be applied in the future across a range of individual and collective problem-solving contexts.

REFERENCES

Amabile, T. (1982). Social Psychology of Creativity: A Consensual Assessment Technique. *Journal of Personality and Social Psychology*, *43*(5), 997–1013.

Amabile T. (1983). The Social Psychology of Creativity: A Component Conceptualisation. *Journal of Personality and Social Psychology*, *45*, 357–376.

Amabile, T. (1996). *Creativity in Context: Update to the Social Psychology of Creativity*. Boulder, CO: Westview.

Amabile, T. (2001). Beyond Talent. *American Psychologist*, *56*(4), 333–336.

Amabile, T., Conti, R., Coon, H., Lazenby, J., & Heroon, M. (1996). Assessing the Work Environment for Creativity. *Academy of Management Journal*, *39*(5), 1154–1184.

Baer, J. G. (2010). Teaching for Creativity in an Era of Content Standards and Accountability. In R. Beghetto, & J. Kaufman (Eds.), *Nurturing Creativity in the Classroom* (pp. 6–19). New York: Cambridge University Press.

Baker, L. M. (2006). Observation: A Complex Research Method. *Library Trends*, *55*(1), 171–189.

Batey, M., & Furnham, A. (2006). Creativity, Intelligence, and Personality: A Critical Review of the Scattered Literature. *Genetic, Social, and General Psychology Monographs*, *132*(4), 355–429.

Bodrova E. L. D. (2007). *Tools of the Mind: "Vygotskian Approach to Early Childhood Education"*. (M. Hall, Ed.) New York : Columbus.

Brinkman, D. (2010). Teaching Creatively and Teaching for Creativity. *Arts Education Policy Review*, 111, 48–50.

Carroll, J. B. (1993), *Human Cognitive Abilities: A Survey of Factor-Analytic Studies*. New York: Cambridge University Press.

Cohen, L. M. (2018). *Research Methods in Education*. London & Abingdon: Routledge.

Davis, G. (2004). *Creativity is Forever*. Dubuque, IA: Kendall Hunt Publishing Company.

Davis, G., Rimm, S., & Siegle, D. (2011). *Education of the Gifted and Talented*. Boston, MA: Pearson Education.

Denscombe, M. (2007). *Good Research Guide*. London: McGraw-Hill Education.

Falk, A., & Szech, N. (2013). Morals and Markets. *Journal of Science*, *340*, 707–711.

Jeffrey, B. & Craft, A. (2004). Teaching Creatively and Teaching for Creativity: Distinctions and Relationships. *Educational Studies*, *30*(1), 77–87.

Karpov, Y. (2014). *Vygotsky for Educators*. New York: Cambridge University Press.

Lloyd, G. (2010). History and Human Nature: Cross-Cultural Universals and Cultural Relatives. *Interdisciplinary Science Reviews, 35*(3–4), 201–214.

Marshall, A., & Batten, S. (2003). Ethical Issues in Cross-cultural Research. *Connections, 3*(1), 139–151.

Patton, M. Q. (2015). *Qualitative Research and Evaluation Methods: Integrating Theory and Practice*. Fourth edition. London: Sage.

Rozesahegyi, T. (2019). Observations. In M. Lambert (Ed.), *Practical Research Methods in Education: An Early Researcher's Critical Guide* (pp. 23–34). London: Taylor & Francis Group.

Skills, T. P. (2011). Framework for 21st Century Learning. *21st Century Education, 3*, 20–50.

Stanley, J. (1991). An Academic Model for Educating the Mathematically Talented. *Gifted Child Quarterly, 35*(1), 36–42.

Sternberg, R. (1985). *Beyond IQ: A Triarchic Theory of Human Intelligence*. New York: Cambridge University Press.

Sternberg, R. (1991). An Investment Theory of Creativity and its Development. *Human Development, 34*, 1–34.

Sternberg, R. (1998). Human Abilities. *Annual Review of Psychology, 49*, 479–502.

Sternberg, R. (2000, June). Patterns of Giftedness: A Triarchic Analysis. *Roeper Review*, 231.

Sternberg, R. (2004). Teaching College Students that Creativity is a Decision. *Guidance & Counseling, 19*(4), 196–200.

Sternberg, R. (2006). The Nature of Creativity. *Creativity Research Journal, 12*, 394–414.

Sternberg, R. (2010). Teaching for Creativity. In R. Beghetto, & J. Kaufman (Eds.), *Nurturing Creativity in the Classroom* (pp. 394–414). New York: Cambridge University Press.

Sternberg, R. K. (2008). *Applied Intelligence*. New York: Cambridge University Press.

Sternberg, R. O. (1997). Creativity as Investment. *California Management Review, 40*, 8–21.

Sternberg, R., & Lubart, T. (1991). An Investment Theory of Creativity and its Development. *Human Development, 34*, 1–32.

Tanggaard, L. (2013). The Sociomateriality of Creativity in Everyday Life. *Culture & Psychology, 19*(1), 20–32.

Torrance, E. (1962). *Guiding Creative Talent*. Englewood Cliffs, NJ: Prentice-Hall Inc.

Torrance, E. (1995). Insights about Creativity: Questioned, Rejected, Ridiculed, Ignored. *Educational Psychology Review, 7*, 313–322.

Torrance, E. P. (1993). Understanding Creativity: Where to Start? *Psychological Inquiry, 4*(3), 232–234.

Treffinger, D. (1988). Components of Creativity: Another Look. *Creative Learning, 2*, 1–4.

Treffinger, D. (1991). Creative Productivity: Understanding its Sources and Nurture. *Illinois Council for Gifted Journal, 10*, 6–8.

Treffinger, D., Schoonover, P., & Selby, E. (2013). *Educating for Creativity and Innovation: A Comprehensive Guide for Research-Based Practices*. Waco, TX: Prufrock Press Inc.

Treffinger, D., Young, G., Selby, E., & Shepardson, C. (2002). *Assessing Creativity: A Guide for Educators*. Storrs, CT: The National Research Center on the Gifted and Talented.
Vygostky, L. S. (2016). *The Psychology of Art* (E. T. Technology, Trans.). London: Cambridge University Press.
Wadaani, M. (2015). Teaching for Creativity as Human Development towards Self-Actualization: The Essence of Authentic Learning and Optimal Growth for All Students. *Creative Education, 6*, 1–10.
Weisberg, R (1999). Creativity and Knowledge. In R Sternberg (Ed.), *Handbook of Creativity* (pp. 226–250). Cambridge: Cambridge University Press.
Woods, P. (1990). *Teacher Skills and Strategies*. London: Falmer.
Woods, P. (1993). *Critical Events in Teaching and Learning*. London: Falmer.
Woods, P. (1995). *Creative Teachers in Primary Schools*. Buckingham: Open University Press.

Creativity and Human Development towards Self-Actualization

Majed Wadaani

5.1 INTRODUCTION

Literature includes a wide range of recommendations that emphasize the importance of allowing youth to use their intellectual potentials at early levels of education. Stanley (1991), for instance, indicated that special efforts at university levels often are not effective with students who have experienced continued earlier boredom and frustration at early ages. Furthermore, creativity is viewed broadly in major theories as related to human attitudes towards life and development that appear more obvious and easier to nurture in young children than it is in older children and adults who have been affected by environments that encourage intellectual conformity, suppress creativity, and overlook talent potential (e.g., Beghetto & Kaufman, 2007; Sternberg, Kaufman, & Grigorenko, 2008; Torrance, 1995).

Sternberg and Kaufman (1998) also indicated that children's multiple abilities will not be utilized if teaching and evaluation systems undervalue creative and practical abilities. As such, teaching should not only be to help students learn facts and think critically about them; teaching should also be for nurturing creative thinking, and facilitating the overall development of students to become the mature adults they are capable of being (Sternberg, 2004). Sternberg (2006) explained that "teaching for creative as well as analytical and practical thinking combined enables children to capitalize on their strengths and to correct or to compensate for their weaknesses in order to be successfully intelligent individuals" (p. 94).

Creativity has been viewed as an important multifaceted phenomenon that can be developed for all students in different areas towards different levels (Davis, 2004). Treffinger, Young, Selby, and Shepardson (2002) stated that "creativity can be expressed in a nearly infinite number of ways in human behavior and has its origins in several components of individual and social experience" (p. 5). Plucker, Beghetto, and Dow (2004) added that "creativity is an important component of problem-solving, healthy social and emotional well-being, and scholastic and adult success" (p. 83). Therefore, teachers should not only teach creatively but also teach for creativity; in order to motivate students to think effectively and become continued creative learners who can make well-informed critical decisions and choices in unexpected situations (Brinkman, 2010; Sriraman, Yaftian, & Lee, 2011; Sternberg, 2010; Torrance, 1995). Miligram and Hong (2009) stated that "societies that do not make every effort to assure that the potential talents of young people are utilized are losing their most valuable natural resource; human capital" p. (161). Accordingly, efforts for education reforms should emphasize teaching for creativity as philosophy of teaching that facilitates human development and self-actualization for all students. Moreover, researchers, educational policymakers, and leaders are supposed to constantly contribute to enhancing school trends and teachers' attitudes towards teaching for creativity utilizing a broad conception of creativity and internalizing positive beliefs about student capability for success.

5.2 CREATIVITY AS A HUMAN PHENOMENON

Numerous attempts for many years have been directed to describe the construct and the principles of creativity which has resulted not only in different constructs of creativity but also in different levels of creativity (Davis, Rimm, & Siegle, 2011). Some of these different views are general and others overlap with other concepts such as giftedness, talent, and intelligence. The major four strands for inquiry on creativity were identified first by Rhodes (1961): the person, process, product, and the press (place/ environment), known collectively as "The 4Ps model". These strands have been the focus of the majority of creativity definitions with different types of interactions and weights.

Humanistic psychologists Maslow (1943, 1968) and Rogers (1954) related creativity to self-actualization as a high-level personal need that requires prior fulfillment of other basic needs including physiological, safety, social, and esteem needs. According to Maslow (1943), self-actualization refers to "the desire for self-fulfillment and being everything that one is capable of becoming" (p. 382). Rogers (1954) indicated that creativity emerges from the need of self-actualization that involves prerequisite personal and environmental conditions that support an internal locus of evaluation, feeling of worth, and freedom of expression. Maslow (1943) described that although creative behaviour has multiple determinations, the products of creative people who are satisfied in their basic needs can be distinguished from the products of unsatisfied creative others.

Although it was not effectively utilized until decades later, the humanistic approach to understanding creativity has been prominent as the essence of broad contemporary conceptions of creativity. Davis (2004) pointed out that the humanistic approach to creativity, through its relationship with self-actualization, provided the most influential concepts in the field of creativity; he summarized that Maslow and Rogers' theories of creativity indicate that the creative person is "a self-actualizing human being who is mentally healthy, self-accepting, democratic minded, fully functioning, and forward growing using all of his/her talents to become what he/she is capable of becoming" (p. 2).

Guilford (1950, 1966) discussed creativity with more emphasis on thinking processes especially divergent thinking in problem-solving. According to Guilford (1950), creativity

> represents patterns of primary abilities that can vary with different spheres of creative activity, and are based in multiple intellectual factors including sensitivity to problems, ideational fluency, flexibility of set, ideational novelty, synthesizing ability, analyzing ability, reorganizing or redefining ability, span of ideational structure, and evaluating ability.
>
> *(p. 454)*

Guilford (1966) added that actual creative performance depends on multiple qualities and dimensions related to potentiality and on what the operating situation allows. The factor of creative potential was central in Guilford's view of creativity, as he (1966) indicated that it plays a significant role, and can be defined as "what an individual brings to a possible creative performance because of his personality structure" (p. 186).

Torrance (1962, 1993) also emphasized on creative thinking process; he indicated that creativity is an important natural process that is based on human needs, which leads to effective learning and continued growth. He (1993) described creative thinking as "the process of sensing of difficulties, problems, gaps in information, missing elements, something askew; making guesses and formulating hypotheses about these deficiencies; evaluating and testing these guesses and hypotheses; possibly revising and retesting them; and, last, communicating the results" (p. 233). According to Torrance's view (1962, 1993), creativity is the essence of scientific discoveries and inventions, and it is also in the realm of everyday living as it is not only reserved for ethereal achieved heights of creation.

Amabile (1982, 1983, 2001) discussed creativity as a dynamic process influenced significantly by the social environment factor. According to Amabile (1982), creativity can be regarded as "the quality of response, process, or products judged to be creative by appropriate observers familiar with the domain in which the response was articulated, the process implemented, or the product created" (p. 1001). She (1983a, 2001) criticized theories of creativity that overemphasize personal talent as a premier source of individual creativity; and proposed her componential model that emphasizes hard work and passionate desire as factors that play central roles in creativity performance.

Amabile's componential model of creativity (1983) includes three basic intra-individual components. The first component is domain-relevant skills/expertise, which represents competencies and talents applicable to the domain or domains in which the individual is working. The second component is creativity-relevant processes, which represent personality characteristics, cognitive styles, and work habits that promote creativity in any domain. The third component is the intrinsic task motivation, the internally driven involvement in the task at hand, which can be influenced significantly by the social environment as an extra individual factor of creativity and innovation. Amabile and her colleagues (1996) added that innovation does not depend only on individual creative ideas; it also requires creative ideas generated and matured by work teams within organizations that support successful implementation. Schools as educational organizations represent environments within which student creative thinking should be nurtured and ideas for innovative products should be supported.

Treffinger (1988, 1991) proposed the COCO model of creativity in which he indicated that creative productivity arises from dynamic interactions among four essential components: characteristics, operations, context, and outcomes. Characteristics include generating ideas, thinking deeper for more ideas, openness and courage to explore ideas, and listening to one's inner voice. Operations involve the strategies and techniques people employ to generate and analyse ideas, solve problems, make decisions, and manage their thinking. Context includes the culture, climate, situational dynamics such as communication and collaboration, and the physical environment in which one is operating. Outcomes are the products and ideas that result from people's efforts.

Sternberg and Lubart (1991, 1996) analysed creativity as an investment process; their Investment Theory of Creativity is a confluence theory according to which creative people are those who are willing and able to "buy low and sell high" in the realm of ideas. Sternberg (2006) described that

> buying low means pursuing ideas that are unknown or out of favor but that have growth potential; and, often, when these ideas are first presented, they encounter resistance; the creative individual persists in the face of this resistance and eventually sells high, moving on to the next new or unpopular idea.
>
> (p. 87)

According to the Investment Theory of Creativity (Sternberg & Lubart, 1991, 1996), creativity requires a confluence of six distinct but interrelated resources: knowledge, intellectual abilities, styles of thinking, personality, motivation, and environment. Knowledge has been considered essential to produce original work and to go beyond what has been already known; it is also necessary because creativity can be domain specific. The intellectual abilities include synthetic, analytic, and practical abilities; using these three abilities, creative individuals can see connections, redefine problems, analyse ideas and judge their potential return, and present ideas in ways that show their values and get accepted for implementation. The styles of thinking represent the ways people prefer to use their intellectual abilities such as the inventing legislative, implementing, and evaluating style.

Sternberg, O'Hara, and Lubart (1997) indicated that everyone possesses every style to some degree, but individuals who want to be creative have to prefer and strengthen the inventing style of thinking which means doing things in novel ways. In addition to knowledge, ability, and style of thinking, individuals also need to have motivation for creativity in order to cope with difficulties faced and move forward with enjoyment. Creativity also requires a self-determined and risk-taking personality that persists for achievement, as well as an environment that supports the investment of ideas and spreads the risks.

Plucker, Beghetto, and Dow (2004) described creativity as "the interaction among aptitude, process, and environment by which an individual or group produces a perceptible product that is both novel and useful as defined within a social context" (p. 90). Beghetto and Kaufman (2007) highlighted the relationship between learning and creativity; they indicated that "the interpretive and transformative process of information is a creative endeavor" (p. 73). They also pointed out that researchers and educators should broaden their conceptions of creativity, and explore how to "best support a lifetime of creative learning and expression" (p. 78). Tanggaard (2013) added that creativity should be viewed as "an everyday phenomenon resulting in continual processes of making the world" (p. 20). Therefore, it can be summarized that creativity can be expressed through different life skills of multiple aspects including the intellectual personal, environmental social, and innovative productivity aspect.

Davis, Rimm, and Siegle (2011) concluded that there are many intellectual abilities that contribute to creative potential; they described the major abilities of creativity that have appeared in creativity literature, as the following:

- Fluency: The ability to produce many ideas in response to an open-ended problem or question, either verbal or nonverbal ones.
- Flexibility: The ability to make different approaches to a problem, think of ideas in different categories, or view a situation from several perspectives.
- Originality: statistical rarity or uniqueness and nonconformity.
- Elaboration: The ability to add details, develop, and implement a given idea.

Davis, Rimm, and Siegle (2011) added that creativity is not limited to these common four abilities. They indicated that other important creative abilities include problem finding, problem sensitivity, problem defining, visualization, analogical thinking, evaluation, intuition, curiosity, independence, resisting premature closure, risk-taking, logical thinking, seeing structure in chaos, discovering relationships, planning, prioritizing, and making good decisions. Literature also indicates that all individuals are capable of enjoying creative thoughts and production, but they may function at different levels of creativity; the following levels of creativity were adopted by Wilson (n.d.):

- Intuitive level: Creative expression for the intrinsic joy of creativity.
- Academic and technical level: Adding power to the creative expression by learning the techniques and skills related to the creative work.
- Inventive level: Going beyond skills and challenging the boundaries to practice untraditional experiments.

- Innovative level: Originality and out of the ordinary productions or ideas that have a guiding academic foundation.
- Genius level: The uniqueness of the ideas or the accomplishments that might have additional genetic aspects.

5.3 CREATIVITY AND HUMAN INTELLIGENCE

Creativity has been discussed in relation to intelligence; however, literature indicates that "intelligence, as like as creativity, lacks a solid operational definitional foundation; definitions of intelligence still range from a neural efficiency perspective to the ability to adapt the self to the environment" (Batey & Furnham, 2006, p. 364). Brody (2000) stated also that "contemporary theorists have not attained consensus about the definition of intelligence" (p. 30). Psychometric approaches of studying intelligence, in turn, have not provided a commonly accepted conclusion about its relationship with creativity (Jauk et al., 2013).

Factor analysis of human abilities provided different views of the quantity and quality of the factors accounting for intelligence components. The view of the single general ability factor "g" that accounts for most human cognitive abilities has been presented earlier, criticized, utilized, and developed (e.g., Spearman, 1904, as cited in Brody, 2000). Guilford's (1956) multiple factors of intelligence and Cattel's (1963) theory of the fluid and crystallized intelligence provoked the field for more investigations and broader multi-aspect views and hierarchical modelling. The study of Flynn (1984) on intelligence development through generations added more complexity and new directions towards understanding and developing intelligence; as he proposed that the environment, Flynn's effect, positively affects human intelligence as it develops. Flynn (2007) concluded that intelligence should be conceptualized like "the atom with multiple components that are held/blended together by the general intelligence factor, and smashed/splitted by the Flynn effect/ environment effect on IQ gains over time" (p. 4).

In regard to creativity, broad theories of intelligence viewed creativity as an aspect and an outcome of intelligence as authors discussed intelligence to be a modifiable phenomenon with multiple factors (e.g., Gardner, 1983; Guilford, 1956; Sternberg, 1985). Guilford (1950) related creativity to divergent thinking process and included it as an aspect in his model of intelligence. In his model of intelligence, the Structure of Intellect, Guilford (1956) criticized the view of the single "g" factor of intelligence and discussed a system of multiple factors of human intellectual abilities that were categorized under general headings including cognition (discovery), production (convergent and divergent thinking), and evaluation. Guilford also contributed to directing research to view intelligence as a multi-aspect phenomenon, as he (1956) indicated that "specifying a number of intelligences would be helpful in more understanding of human abilities" (p. 291).

Gardner (1983, 1999) was against the notion of determining human intelligence with a single general factor. He has a wide view of intelligence that considers creativity as a parallel ability, with a focus on providing the conditions for the utilization of human multiple intelligences rather than just concerning about assessment. According to Gardner (1983), each human being possesses a blend, with different levels, of several basic intellectual competencies/semi-independent intelligences that include linguistic, logical-mathematical, musical, bodily kinesthetic, spatial, interpersonal, and intrapersonal intelligence. Gardner (1999) also discussed the possible existence of three additional kinds of intelligence, the naturalist, spiritual, and the existential intelligence. He also indicated that the discussion is still open for additional types of intelligences as human intelligence is hard to capture.

According to Gardner (1999), creativity is a part of the intellectual realm that is parallel with, but different from intelligence. Gardner (1999) defined creativity as "the faculty of solving problems, creating products, or raising issues in a domain in a way that is initially novel, but is eventually accepted in on or more cultural setting" (p. 116). He (1999) pointed out that the major difference between creativity and intelligence is that creativity is a domain-specific activity that results in novel products or changes in the domain. Gardner (1999) admitted that the relationship between intelligence and creativity is complex, as he discussed different possible factors, kinds, and levels of creativity that emerged from personality theories and social psychology.

Sternberg's Triarchic Theory of Intelligence (1985) was one of theories that clearly included creativity as an aspect of intelligence. Sternberg's Triarchic Theory of Intelligence (1985) viewed creativity as an aspect among three interacting aspects of intelligence. The first aspect is analytical intelligence which involves information processing skills. The second is practical intelligence which involves using mental components of intelligence to adapt to, shape, or select environment that is appropriate for oneself. The third aspect is creative intelligence which involves using mental components of intelligence to create new products or make new discoveries.

In terms of empirical research, the issue of whether current Intelligence Quotient (IQ) tests truly assess human intelligence, in addition to the difficulties in the assessment of creativity itself, has been the major challenge towards providing a commonly accepted conclusion about the relationship between human intelligence and creativity. Even with the reliance on IQ tests, Torrance (1993) argued that the possession of high intelligence as measured by IQ tests is not enough for outstanding creative success. He explained that creative thinking includes responding constructively to existing or new situations which may take time for incubation, rather than merely adapting to them using limited intellectual abilities.

Jauk et al. (2013) stated that "investigations of the relationship between intelligence and creative potential provide a scattered view" (p. 214). They added that even studies of the prominent threshold hypothesis of the relationships between intelligence and creativity showed inconsistent results. They indicated that the threshold of 120 IQ points, as the minimum level of intelligence necessary for creativity, represents an educated guess as empirical reliable supporting studies still are inadequate in this aspect.

Batey and Furnham (2006) indicated that making unequivocal conclusions about the relationship between creativity and intelligence is unwise. They indicated that it is unwise because of inconsistent definitions provided for creativity and intelligence, the different types of psychometric instruments involved in the studies, as well as the influential traditional issues faced in such measurement studies including the design of IQ and creativity tests, sample sizes, and statistical analyses. Kim (2005) concluded that "the negligible relationship between creativity and IQ scores indicates that even students with low IQ scores can be creative" (p. 65).

In conclusion, data obtained from empirical studies of the relationship between creativity and intelligence as measured by IQ tests reveal that both high IQ and average students have the potential to develop and expand their creative thinking skills (Kim, 2005). Research in this aspect also indicates that creativity is not solely dependent upon intelligence as studies showed that average students can score higher than high IQ students on some parts of creativity tests, as well as high IQ students are not always able to score consistently on all tasks of cognitive problem-solving and creativity tests (Russo, 2004).

5.4 CREATIVITY AS HUMAN DEVELOPMENT TOWARDS SELF-ACTUALIZATION

Based upon what has been discussed in relevant literature about creativity, it is practical in the field of education that creativity is perceived as "Human

Development towards Self-Actualization" based upon Maslow's (1943) definition of "Self-actualization" which refers to "the desire for self-fulfillment and being everything that one is capable of becoming" (p. 382). The humanistic self-actualization approach to creativity (Maslow, 1943, 1968; Rogers, 1954) has been supported either explicitly or implicitly by many scholars. Davis (2004) admitted that the relationship between creativity and self-actualization is one of the most influential concepts in the field of creativity and argued that creativity is a way of thinking and living that leads to personal development and a productive successful life.

Creativity is a process of growth and the ultimate outcome of continued education. It represents ongoing overall development of humans/students that leads them to positively actualize themselves to reach the highest level possible appropriate for their special aptitudes, potentials, circumstances, and community needs. As the highest level of self-actualization may vary from an individual to another, students should be facilitated to develop positive belief systems related to their personal abilities, achievement goals, and future success, and lead them to keep an active and regenerated desire for self-actualization through unlimited high-level personal goals of achievement. Creativity, as human development, includes a broad range of personality characteristics, life skills, general mental skills, and domain-related skills (specialized creativity). Creativity as a process of thinking and production is included as an essential part of the 21st-century skills necessary for success in such a changing world (The Partnership for 21st Century Skills, 2011); however, creativity in its comprehensive meaning actually represents all reported skills of the 21st century including learning and innovation skills, information media and technology, life, and career skills.

Moreover, creativity is not limited to particular fields or exclusive to those who develop original ideas or products in such domains. Creativity is impeded in our daily life; we practice creativity explicitly and implicitly dealing with what we face every day, but at different levels. Creativity can be as simple as the feeling of being creative, the effective utilization of intellectual abilities, the successful personal and social living, the expression of unique performance in a special area, or it can go beyond the use of basic intellectual abilities by capitalizing on advanced skills for the development of original ideas and products that are innovative in a way that modifies or adds to the existing body of knowledge. Therefore, creativity is the crystalized aspect of human intelligence with its complex components and processes, not measured only by IQ tests; creativity is also the product of applied human intelligence with its multiple life dimensions, not measured only by standardized tests. The most successful creative person is the one who keeps positive beliefs and a growth mindset (Dweck, 2006), and works on continued development processes of his/her abilities to reach the highest level possible, so he/she can think and act creatively at a high level in all life aspects related to personality (cognitive, affective, and behavioural traits), environment, and productivity.

The academic aspect is significant in student life success. Demonstrating evolving skills related to a certain academic field can be a sign of creativity in this field, mathematical creativity for example. Signs of creativity in mathematics can be further developed in order to reach the highest level of creativity which represents the production of novel solutions that extend knowledge in the field. For students to achieve self-actualization, they need to be as creative as possible on both sides; general/personal/life creativity, and domain-specific creativity, considering that the absence of original/novel production (innovation) does not mean the absence of creativity as a way of thinking and living.

Furthermore, it is not enough that students be successful in such academic subjects, and even to be eminent in a certain field of study or profession, without having developed personal and social skills that allow them to employ and direct their skills into the right path for personal and community development. Teaching in schools should be administered to lead all students to be creative personally, socially, and academically with the same weight of importance, as interrelated factors of authentic achievement and success.

Teaching for creativity applies to all areas of study because personal and social skills of the successful life (general creativity) should be developed explicitly and implicitly through each subject class as an important educational endeavour for the academic skills of each subject (domain-specific/specialized creativity). Creativity for elementary school-age children should not be evaluated compared to levels of creativity of individuals at older ages. Signs of creativity in early ages might be simple and not worthy for some adult observers, but they are actually significant when they are evaluated compared to the abilities of peers at the same age, and can be further developed for promising results. Children's emerging abilities should be fostered through teaching for creativity which is the real achievement; creativity as real achievement means human development personally, socially, and academically towards self-actualization, i.e. self-actualization as a central ongoing human need.

5.5 TEACHING FOR CREATIVITY

Creativity is not a learning objective to be achieved and measured, but it is a personal component that all individuals should keep developing to the highest level possible. Teaching for creativity is not a teaching method, but it is a teaching philosophy that all teachers should adopt in order to facilitate the continued development of human resources. Teachers should teach for creativity as a crystallized modifiable human capability with complex dimensions and an endless ceiling. Teaching for creativity is a philosophy through which multiple teaching methods can be modified or generated to lead students to develop their creativity in different contexts towards different levels. Most teaching methods that are considered in literature as effective methods in increasing student academic achievement can also be effective in developing student creativity if teachers' beliefs about teaching and achievement change.

When teachers understand their roles as a facilitator of optimal human development and internalize positive beliefs about creativity as "the hub of real achievement" (Forster, 2012, p. 281), they can teach for creativity, as they will engage in group and self-learning for professional development that leads them to apply best practices. Teachers can practice some teaching styles to create an environment of creativity that strengthens the teaching methods that they typically use. They may also use some additional teaching techniques to enhance creativity for all students, or they may design teaching methods to use specifically for fostering certain aspects of creativity.

Literature includes several teaching techniques, styles, methods, and models that have been recommended to develop creativity for all students. Gifted and talented students have been viewed to have high creativity potential, and as such, they need additional special care for maximizing their potential high creativity. Several models have been proposed specifically for developing certain thinking skills; however, some of these models have an emphasis on just some components of creativity such as divergent thinking and original products. Creativity is not only about divergent thinking; it is a construct that includes all aspects of human life that should be nurtured implicitly and explicitly in all contexts utilizing all available opportunities and resources for the overall optimal development of personality and social skills.

Nickerson (2010) pointed out that teaching for creativity can be carried out simply by avoiding negative beliefs and practices that discourage creativity such as that are related to authority, ownership of ideas, making mistakes, questioning, tests, intelligence, originality, and student capability for development and success. He added that teaching for creativity is based on helping students to establish positive beliefs about their abilities and strengthen their attitudes towards creativity as a life skill. Therefore, teachers can develop student creativity even without the expenditure of extra time or the introduction of new curriculum.

Baer and Garrett (2010) pointed out that teaching for creativity and teaching specific content knowledge are not in opposition; they added that "teachers can successfully meet accountability standards and promote creativity in their classrooms" (p. 19). Treffinger, Schoonover, and Selby (2013) indicated that teachers, who hold a belief that student creativity can and should be developed, can teach for creativity even if they face challenges and concerns related to the educational system. For example, they stated that "creativity and the U.S. Curriculum Common Core State Standards (CCSS) are not incompatible when approached effectively and some assumptions change such as those related to the standardized tests as being the only effective way of evaluation" (p. 254). Sheffield and her colleagues (2013) agreed that the CCSS for mathematics can be implemented in various ways to provide opportunities for developing creativity and nurturing advanced learners. Therefore, any teacher with positive beliefs and continued professional learning can construct learning activities that effectively integrate curriculum standards and tools for creative thinking (Sheffield et al., 2013; Treffinger, Schoonover, & Selby, 2013).

Teaching for creativity is based on providing a positive school environment. Snyder, Lopez, and Pedrootti (2011) emphasized on the importance that teachers interact positively with students in order to identify and expand their strengths. Positive schooling, according to Snyder, Lopes, and Pedrotti (2011), represents

> an approach to education that consists of a foundation of care, trust, and respect for diversity, where teachers develop tailored goals for each student to engender learning and work with him or her to develop the plans and motivation to reach their goals; positive schooling includes the agendas of installing hope in students and contributing to the larger society.
>
> *(p. 415)*

The positive school environment that enhances student creativity requires that teachers provide a psychologically safe and motivating climate (Amabile, 1996; Rogers, 1954) that allows each student to think, try, share, use different ways, make mistakes, question, feel worthy, build autonomy, and achieve self-esteem. Psychological safety can be achieved by accepting and valuing all students' contributions, encouraging participation and collaboration, limiting competition, avoiding punitive assessment tests, and contributing to addressing personal and social student issues such as bullying, taunting, home and social status (Fairweather & Cramond, 2010). Metcalf (2010) added that student feeling of safe is influenced by the language that teachers use, and the way in which student academic issues are addressed. She suggested solution-focused teaching to help students take responsibility for their issues, focus on solutions, and find opportunities for further development. Solution-focused teaching is based on the teacher as a facilitator of student success, by helping them externalize negative assumptions and internalize positive beliefs about their personalities, abilities, and responsibility for learning which is essential for creativity development.

Amabile and her colleagues (1996) believe that creativity can be encouraged through factors that promote intrinsic motivation with a positive sense of challenge and a focus on the work itself. Such important factors of creativity development that they declared include autonomy space, creativity encouragement, commitments with clear goals, mutual openness to ideas, and constructive challenge with appropriate reactions and feedback. Amabile (2001) added that policymakers and educators should not worry whether students have the talent required to be provided with special education to achieve creative performance; rather, they should focus on creating opportunities for all individuals to master effective skills that help them develop their expertise in the domain and think creatively within environments that support active, deep engagement with challenging work, remembering that "creativity depends not only on brilliance and wit but also on discipline and passionate desire" (p. 335).

Sternberg (2010) indicated that creativity is "a habit that can be encouraged" (p. 394); he discussed essential principles of teaching for creativity. According to Sternberg (2010), the main components that prompt creativity include opportunities to engage in it, encouragement when students take advantage of these opportunities, and rewards when students think and behave creatively responding to such encouragement. He added that teaching for creativity requires teachers to be models of creative thinkers and performers for the purpose of lasting impacts on students' attitudes towards creativity.

The supportive, rewarding school environment of creative thinking, according to Sternberg (2010), entails using constructive practices of teaching and evaluation that allow space for creative thinking and behaviours, as creativity would be discouraged by the inappropriate use of conventional standardized tests that lead to conformity in both teachers' practices and student thinking. Some of the practices that he recommended to promote creativity include encouraging students to define and redefine their own problems and choices, question and analyse assumptions, find and work on what they love to do, practice presentation of their ideas and defending their positions, value knowledge and be continued learners, think across disciplines, identify and address obstacles, know about and appreciate creative individuals' lives and contributions, tolerate ambiguity, collaborate with others, take intellectual risks, learn from mistakes, and build self-efficacy.

Baer and Garrett (2010) believe that teaching for creativity is not only about developing divergent thinking. They explained that teaching for creativity is based on the balanced use of several effective theories and approaches of learning and teaching including the balanced utilization of intrinsic and extrinsic motivation, student-centred and teacher-centred, constructionist and transmissive ways of learning. For example, Amabile's Social Psychology Theory of Creativity (1983, 1996) suggested that creativity is enhanced when students work in activities that are intrinsically motivating, "and it is in decrease when students work only for extrinsic rewards". As the balanced use of different approaches of teaching and learning is recommended, Baer and Garrett (2010) indicated that teachers can develop student creativity by giving more emphasis in activities that increase intrinsic motivation at certain times, and allow external rewards at other times, in order to capitalize on both approaches, and meet the curriculum standards of the educational system. External rewards being simple and unexpected are important conditions to ensure that intrinsic motivation is not affected negatively. Based on previous review of literature, creativity can be enhanced for all students using different methods and techniques; the major factor of successful teaching for creativity is that teachers internalize positive beliefs and student capability for success, and hold positive attitudes towards teaching for creativity in order to facilitate optimal growth for all students.

5.6 CONCLUSION

In conclusion, teaching for creativity is important to be adopted by teachers as philosophy of teaching in order to facilitate student overall development and self-actualization and thus contributing to nurturing future creative leaders who can positively affect their communities. The essence of teaching for creativity is providing a supportive learning environment that makes students feel safe, worthy, and encouraged to express their natural curiosity and abilities, with adequate support for developing productivity, personality, and social skills. These conditions are fundamentals for extending student academic abilities and developing domain-specific creativity and talent. In order for teachers to teach for creativity, professional training and other features of support should be adequately available. Educational policy and strategic programs should be developed in light of teachers' needs in a way that ensures prior healthy conditions for successful implementation; this includes a clear mission and vision, continuous professional development, convinced satisfied staff, sufficient facilities, and a cooperative safe school climate that supports creativity.

REFERENCES

Amabile, T. (1982). Social Psychology of Creativity: A Consensual Assessment Technique. *Journal of Personality and Social Psychology, 43*(5), 997–1013.

Amabile, T. (1983). The Social Psychology of Creativity: A Componential Conceptualization. *Journal of Personality and Social Psychology*, *45*(2), 357–376.

Amabile, T. (1996). *Creativity in Context: Update to the Social Psychology of Creativity*. Boulder, CO: Westview.

Amabile, T. (2001). Beyond Talent. *American Psychologist*, *56*(4), 333–336.

Amabile, T., Conti, R., Coon, H., Lazenby, J., & Heroon, M. (1996). Assessing the Work Environment for Creativity. *Academy of Management Journal*, *39*(5), 1154–11841.

Baer, J., & Garrett, T. (2010). Teaching for Creativity in an Era of Content Standards and Accountability. In R. Beghetto, & J. Kaufman (Eds.), *Nurturing Creativity in the Classroom* (pp. 6–19). New York: Cambridge University Press.

Batey, M., & Furnham, A. (2006). Creativity, Intelligence, and Personality: A Critical Review of the Scattered Literature. *Genetic, Social, and General Psychology Monographs*, *132*(4), 355–429.

Beghetto, R., & Kaufman, J. (2007). Toward a Broader Conception of Creativity: A Case for "mini-c" Creativity. *Psychology of Aesthetics, Creativity, and the Arts*, *1*(2), 73–79.

Brinkman, D. (2010). Teaching Creatively and Teaching for Creativity. *Arts Education Policy Review*, *111*, 48–50.

Brody, N. (2000). History of Theories and Measurements of Intelligence. In S. Robert (Ed.), *Handbook of Intelligence* (pp. 16–33). Cambridge: Cambridge University Press.

Cattel, R. (1963). Theory of Fluid and Crystallized Intelligence: A Critical Experiment. *Journal of Educational Psychology*, *54*(1), 1–22.

Davis, G. (2004). *Creativity Is For Ever.* Dubuque, IA: Kendall Hunt Publishing Company.

Davis, G., Rimm, S., & Siegle, D. (2011). *Education of the Gifted and Talented.* Boston, MA: Pearson Education.

Dweck, C. (2006). *Mindset: The New Psychology of Success.* New York: Random House.

Fairweather, E., & Cramond, B. (2010). Infusing Creative and Critical Thinking into the Curriculum Together. In R. Beghetto & J. Kaufman (Eds.), Nurturing Creativity in the Classroom (pp. 113–141). Cambridge: Cambridge University Press. doi:10.1017/CBO9780511781629.007

Flynn, J. (1984). The Mean IQ of Americans: Massive Gains 1932 to 1978. *Psychological Bulletin*, *95*(1), 29–51.

Flynn, J. (2007). *What Is Intelligence? Beyond the Flynn Effect.* New York: Cambridge University Press.

Forster, J. (2012). Creativity: The Hub of Real Achievement. *Gifted Education International*, *28*(3), 281–299.

Gardner, H. (1983). *Frames of Mind: The Theory of Multiple Intelligences.* New York: Basic Books.

Gardner, H. (1999). *Intelligence Reframed. Multiple Intelligences for the 21st Century.* New York: Basic Books.

Guilford, J. (1950). Creativity. *The American Psychologist*, *5*(9), 444–454.

Guilford, J. (1956). The Structure of Intellect. *Psychological Bulletin*, *53*(4), 267–293.

Guilford, J. (1966). Measurement and Creativity. *Theory into Practice*, *5*(4), 186–202.

Jauk, E., Benedek, M., Dunst, B., & Neubauer, A. (2013). The Relationship between Intelligence and Creativity: New Support for the Threshold Hypothesis by Means of Empirical Breakpoint Detection. *Intelligence, 41*, 212–221.

Kim, K. (2005). Can only Intelligent People Be Creative? A Meta-Analysis. *The Journal of Secondary Gifted Education, 14*(3), 57–66.

Maslow, A. (1943). A Theory of Human Motivation. *Psychological Review, 50*(4), 370–396.

Maslow, A. (1968). *Toward a Psychology of Being.* New York: Harper.

Metcalf, L. (2010). *Teaching toward Solutions.* Williston, VT: Crown House Publishing.

Miligram, R., & Hong, E. (2009). Talent Loss in Mathematics: Causes and Solutions. In R. Leikin, A. Berman, & B. Koichu (Eds.), *Creativity in Mathematics and the Education of Gifted Students* (pp. 149–161). Rotterdam, The Netherlands: Sense Publishers.

Nickerson, R. (2010). How to Discourage Creative Thinking in the Classroom. In R. Beghetto, & J. Kaufman (Eds.), *Nurturing Creativity in the Classroom* (pp. 1–5). New York: Cambridge University Press.

Plucker, J., Beghetto, R., & Dow, G. (2004). Why isn't Creativity More Important to Educational Psychologists? Potential, Pitfalls, and Future Directions in Creativity Research. *Educational Psychologist*, 39, 83–96.

Rhodes, M. (1961). An Analysis of Creativity. *Phi Delta Kappan*, 42, 305–310.

Rogers, C. (1954). Toward a Theory of Creativity. *ETC: A Review of General Semantics*, 11, 249–260.

Russo, C. F. (2004). A Comparative Study of Creativity and Cognitive Problem-solving Strategies of High-IQ and Average Students. *Gifted Child Quarterly*, 48(3), 179–189.

Sheffield, L., Johnson, S., Tassel-Baska, J., Adama, C., Cotabish, A., & Mursky, C. (2013). *Using the Common Core State Standards for Mathematics with Gifted and Advanced Learners.* Waco, TX: Pruforck Press Inc. with NAGC, NCSM, and NCTM.

Snyder, C., Lopez, S., & Pedrootti, J. (2011). *Positive Psychology: The Scientific and Practical Explorations of Human Strengths.* Thousand Oaks, CA: SAGE Publications, Inc.

Sriraman, B., Yaftian, N., & Lee, K. (2011). Mathematical Creativity and Mathematics Education: A Derivative of Existing Research. In B. Sriraman, & K. Lee (Eds.), *The Elements of Creativity and Giftedness in Mathematics* (pp. 119–130). Rotterdam, The Netherlands: Sense Publishers.

Stanley, J. (1991). An Academic Model for Educating the Mathematically Talented. *Gifted Child Quarterly*, 35(1), 36–42.

Sternberg, R. (1985). *Beyond IQ: A Triarchic Theory of Human Intelligence.* New York: Cambridge University Press.

Sternberg, R. (2004). Teaching College Students that Creativity Is a Decision. *Guidance & Counseling*, 19(4), 196–200.

Sternberg, R. (2006). The Nature of Creativity. *Creativity Research Journal*, 18(1), 87–98.

Sternberg, R. (2010). Teaching for Creativity. In R. Beghetto, & J. Kaufman (Eds.), *Nurturing Creativity in the Classroom* (pp. 394–414). New York: Cambridge University Press.

Sternberg, R., & Kaufman, J. (1998). Human Abilities. *Annual Review Psychology*, 49, 47–502.

Sternberg, R., Kaufman, J., & Grigorenko, E. (2008). *Applied Intelligence*. New York: Cambridge University Press.

Sternberg, R., & Lubart, T. (1991). An Investment Theory of Creativity and its Development. *Human Development*, 34, 1–32.

Sternberg, R., & Lubart, T. (1996). Investing in Creativity. *American Psychologist*, 51(7), 677–688.

Sternberg, R., O'Hara, L., & Lubart, T. (1997). Creativity as Investment. *California Management Review*, 40(1), 8–21.

Tanggaard, L. (2013). The Sociomateriality of Creativity in Everyday Life. *Culture & Psychology*, 19(1), 20–32.

The Partnership for 21st Century Skills. (2011). *Framework for 21st Century Learning*. Washington, DC: Partnership. Retrieved on November 12, 2014 from http://www.p21.org/storage/documents/1.__p21_framework_2-pager.pdf.

Torrance, E. P. (1962). *Guiding Creative Talent*. Englewood Cliffs, NJ: Prentice-Hall, Inc.

Torrance, E. P. (1993). Understanding Creativity: Where to Start? *Psychological Inquiry*, 4(3), 232–234.

Torrance, E. P. (1995). Insights about Creativity: Questioned, Rejected, Ridiculed, Ignored. *Educational Psychology Review*, 7(3), 313–322.

Treffinger, D. (1988). Components of Creativity: Another Look. *Creative Learning*, 2(5), 1–4.

Treffinger, D. (1991). Creative Productivity: Understanding its Sources and Nurture. *Illinois Council for Gifted Journal*, 10, 6–8.

Treffinger, D., Schoonover, P., & Selby, E. (2013). *Educating for Creativity and Innovation: A Comprehensive Guide for Research-Based Practices*. Waco, TX: Prufrock Press Inc.

Treffinger, D., Young, G., Selby, E., & Shepardson, C. (2002). *Assessing Creativity: A Guide for Educators*. Storrs, CT: The National Research Center on the Gifted and Talented.

Wilson, L. (n.d.). *Levels of Creativity*. Retrieved on October 16th, 2014 from http://thesecondprinciple.com/creativity/creativetraits/levels-of-creativity/.

Supporting Creativity and Mathematical Talent Development

Majed Wadaani

6.1 INTRODUCTION

Nations around the world seek to maintain a global leadership position in productivity and scientific development by focusing on youth development. As such, youth, as future leaders for any nation, need to be exposed to educational experiences at an early age that will help them become more independent, analytical, critical, creative, cooperative, and thus personally and socially successful in dealing with real-life problems and needs. These significant skills are very important to be nurtured in general classrooms for all students, with additional enrichment opportunities for those who express high academic abilities and potential giftedness (Wadaani, 2015a).

Allowing youth to use their intellectual potential at early levels of education is highly critical, as special efforts at university levels often are not effective for students who have experienced continued earlier boredom and frustration (Stanley, 1991). Literature also includes that creativity, as an attitude towards life and development, is more obvious and easier to nurture in young children than it is in older children, and adults who have been affected by environments that encourage intellectual conformity suppress creativity and overlook talent potential (e.g., Beghetto & Kaufman, 2007; Sternberg, Kaufman, & Grigorenko, 2008; Torrance, 1995). Therefore, creativity is supposed to be an essential part of each early education curriculum.

Creativity is described as an important multifaceted phenomenon that can be developed for all students in different areas towards different levels (Davis, 2004; Wadaani, 2015b). For example, Treffinger, Young, Selby, and Shepardson (2002) stated that "Creativity can be expressed in a nearly infinite number of ways in human behavior and has its origins in several components of individual and social experience" (p. 5). Plucker, Beghetto, and Dow (2004) added that "creativity is an important component of problem-solving, healthy social and emotional well-being, and scholastic and adult success" (p. 83). Therefore, it is important that creativity should be nurtured for all students, and further supported with gifted students in order to help students become continued creative learners who can make well-informed critical decisions and choices in unexpected situations (Brinkman, 2010; Sriraman, Yaftian, & Lee, 2011; Sternberg, 2010; Torrance, 1995; Wadaani et al., 2016).

On the other hand, giftedness in mathematics has been a global great interest to look for and invest in. The National Council of Teachers of Mathematics (NCTM) (1980) stated that "outstanding mathematical ability is a precious societal resource, sorely needed to maintain leadership in a technological world" (p. 18). Maintaining world influential status and making consistent progress require education systems that effectively nurture creativity and further develop mathematical talent (Mann, 2005), responding to the essential roles that mathematics plays in the development of sciences, technology, economics, and various branches of industry (Leikin, 2009). Accordingly, schools and teachers face a major commitment to respond to the need for nurturing student creativity and further supporting those with potential giftedness in mathematics.

As stated by Miligram and Hong (2009), "societies that do not make every effort to assure that the potential talents of young people are utilized are losing their most valuable natural resource; human capital" (p. 161). Therefore, this research study is a kind of effort that may fill a part of the gap in this issue where the literature is reviewed and supportive information is synthesized for a proposed model designed for understanding and supporting the development of creativity and talent in mathematics hoping to lead to promising effective instructional practices in schools. The proposed model is a developmental model designed in order to open windows through which multiple educational objectives can be achieved. It is based on internalizing education as a process of human development, and mathematics as an important field in scientific development. It also provides foundations for a sequence of teaching focus aspects that are significant not only for those who have potential giftedness but also for all students as the atmosphere of creativity enhances achievement for all.

The proposed model has all students in the first part where the focus is teaching for creativity development (TCD); teaching for creativity enhances authentic learning and facilitates real achievement for all students (Bahar & Maker, 2011; Burleson, 2005; Sheffield, 2013; Wadaani, 2015a, 2015b); creativity here represents overall human development towards self-actualization. Self-actualization as a motive of creativity was defined by Maslow (1943) as "the desire for self-fulfillment and being everything that one is capable of becoming" (p.382). The second part of the model entails teaching for mathematical talent development (TMTD), where students with high potential receive a type of education that extends the efforts of developing creativity in the first part, by further developing mathematical abilities, including mathematical creativity, leading them to invest on their giftedness, become talented and future mathematicians who represent human resources that all nations always seek to attain. The two parts of the model TCD+TMTD connect and function together as one process which gives us the whole model, Teaching for Creativity and Mathematical Talent Development (T-CMTD).

6.2 RESEARCH PROBLEM AND QUESTIONS

Misunderstanding about nurturing creativity and talent development widely exists among teachers (Wadaani, 2015a), and there is still a need for more theoretical foundations and applied models that can assist teachers to get a clear positive understanding and play their roles in human development at schools. In this study, the researcher tried to synthesize literature in creativity and mathematics gifted education to design a proposed model for understanding and supporting hypothesized development phases of creativity and mathematical talent, through which appropriate teaching practices are presented, and requirements for its effective application are discussed based on teachers' perspectives. The major purpose is to enhance school trends and teachers' attitudes towards making creativity and mathematical talent development a high-priority educational goal with a clear understanding of underlying concepts. As such, the current research study is guided by the following questions:

1. What are features of the model that can lead mathematics teachers to better understand and support creativity and mathematical talent development in general education schools?
2. What are the preliminary requirements for the effective application of this model in the actual environments of general education schools?

6.3 SIGNIFICANCE

Gifted individuals mostly are those who contribute a large part to improve many life aspects by solving problems creatively, developing genuine innovations, and leading communities to productivity. Therefore, effective gifted education is an important issue worldwide, and at the centre of this issue is creativity as the most important topic in the education of gifted and talented children (Davis et al., 2011). Moreover, in the current century, mathematics teachers face a major educational commitment to raise student creativity and nurture giftedness in mathematics in order to play the expected role in enhancing society development and scientific productivity. However, the findings of previous studies included that teachers' perceptions and attitudes towards supporting creativity and talent development were not at the desired level with an unclear image of student creativity and giftedness (Wadaani, 2015a).

Accordingly, the importance of this research study appears here as an effort to provide a model that can help teachers understand creativity and giftedness in a way that allows them effectively support talent development for students. This kind of model is theoretically significant because it will move the scientific discussion forward in the field, and it is also practically significant because it provides all teachers with a clear simple image of how promising students can be supported to move through proposed talent development stages.

6.4 METHOD AND PROCEDURES

In order to build a model that facilitates understanding and supporting creativity and mathematical talent development in general education schools, the researcher conducted a deep analysis of relevant literature and previous studies. The major theories about creativity and giftedness are described and connected for common themes. Then, the rationales for a proposed model were extracted and used as bases for building the T-CMTD model. The first version of the model was presented to four specialists in the field of mathematics and gifted education for their feedback; the common recommendations were utilized for improving the outcome and obtaining the final version of the model. After that, a group of mathematics teachers was needed to continue the validation process and get answers for the second question. A convenience sample method was used: the researcher sent invitation messages to accessible teachers from Jazan Region, Saudi Arabia, for voluntarily participation. The researcher had a total of 14 mathematics teachers' agreements for participation. Accordingly, the final version of the model was presented to a group of nine teachers for open discussion; the main purpose of this procedure was to ensure that all terms included in the model are clear, and connections of its components are easily understandable for the target population; the research population here is general education mathematics teachers. Results from the teacher group discussion did not yield any major changes needed in the model.

To obtain data for the second question of this research study, a focus group interview was conducted with other five mathematics teachers from the sample who are not included in the previous validity procedure. The focus group interview aimed at getting teachers' perspectives about the preliminary requirements for effective utilization of the proposed model in the current environments of general education schools. A descriptive introduction was provided; then, the focus group interview started around the following major question: what are the preliminary requirements for effective application of the T-CMTD model in the actual environments of general education schools? Other follow-up and open-ended questions were utilized based on the presented feedback to have teachers deeply reflect on the application requirements of the proposed model and share their perspectives in a comfortable situation. Qualitative data from the focus group interview were transcribed and analysed to find patterns and themes; then, common elements were identified that represent teachers' perspectives about the preliminary requirements for the effective utilization of the proposed model.

6.5 REVIEW OF LITERATURE

The NCTM (2000) pointed out that "in this changing world, those who understand and master mathematical abilities will have significantly enhanced opportunities and options for shaping a productive future" (p. 5). Providing students with an environment that develops their mathematical abilities and further nurtures creativity in the domain should be a primary goal for educational systems that seek to foster creative leaders of societies who can contribute in making life better. However, current dominating practices that are based on providing special enrichment programs only outside the classroom are not adequate; as creativity should be nurtured for all students, such special enrichment programs are also hindered by identification procedures that might overlook some gifted students due to the ambiguous nature of creativity and giftedness (Wadaani, 2015a).

Literature includes different theories and principles of creativity which have resulted in different constructs, factors, and levels (Davis, Rimm, & Siegle, 2011; Rhodes, 1961). Some of these theories provided a narrow restricted perceptive, and others discussed creativity in a broad view (Wadaani et al., 2016). Creativity has been defined as a lifestyle, as a factor of effective learning, as a component of giftedness, as a separate category of talent, and as original productivity (Wadaani, 2015b). Plucker, Beghetto, and Dow (2004), for example, described creativity as "the interaction among aptitude, process, and environment by which an individual or group produces a perceptible product that is both novel and useful as defined within a social context" (p. 90). Creativity is also included in dominating broad theories of intelligence where creativity was viewed as an aspect and as an outcome of intelligence (e.g., Gardner, 1983; Sternberg, 1985).

One of the prominent approaches to understanding creativity is the humanistic approach, which has been influential as the essence of broad contemporary conceptions of creativity (Davis, 2004). According to Maslow (1943, 1968) and Rogers (1954), creativity is related to self-actualization as a high-level personal need that requires prior fulfillment of other basic needs like physiological, safety, social, and esteem needs. Maslow (1943) pointed out that self-actualization refers to "the desire for self-fulfillment and being everything that one is capable of becoming" (p. 382). Rogers (1954) added that the need of self-actualization is the motive of creativity; self-actualization, however, involves prerequisite personal and environmental conditions that support an internal locus of evaluation, feeling of worth, and freedom of expression. Maslow (1943) further described that creative behaviours and products are affected by the satisfaction of basic needs. Accordingly, Maslow and Rogers' theories of creativity suggest that the creative person is "a self-actualizing human being who is mentally healthy, self-accepting, democratic minded, fully functioning, and forward growing using all of his/her talents to become what he/she is capable of becoming" (Davis, 2004, p. 2).

Based on previous foundations, creativity can be described as a multifaceted human development phenomenon that includes a broad range of personality characteristics, life skills, general mental skills, and domain-related skills (Wadaani, 2015b). Demonstrating evolving skills related to a certain academic domain can be a sign of creativity in that domain, mathematical creativity for example. Signs of creativity in mathematics can be further developed in order to reach the highest level of creativity which represents the production of novel solutions that extend knowledge in the domain.

Creativity in mathematics also lacks a common definition; however, mathematicians agree that creativity is a major element of mathematical activities (Lee, Hwang, & Seo, 2003; Yuan & Sriraman, 2011). Contemporary mathematics educators view creativity as "an orientation or disposition toward mathematical activity that can be fostered broadly in the general school population" (Silver, 1997, p. 75). Mann (2006) pointed out that creativity is the essence of mathematics and the most influential factor of effective learning, mathematical talent development, and future mathematical accomplishments. Leikin (2009) also indicated that mathematical creativity is "the dynamic property of the human mind and each child's creative potential can be developed and realized, or on the contrary, deprived" (p.129).

Sriraman (2005) added that although creativity in mathematics is often looked at as the exclusive domain of professional mathematicians, "students at the general education level are capable of creativity" (p.86). He explained that at the professional level, mathematical creativity can be defined as the ability to produce original work that significantly extends the body of knowledge, and/or one that opens up avenues of new questions for other mathematicians. However, mathematical creativity in school settings can be defined as

> the process that results in unusual and/or insightful solutions to a given problem or analogy problems; and/or the formulation of new questions and/ or possibilities that allow an old problem to be regarded from a new angle requiring imagination which is similar to those for creativity in professional mathematics.
>
> (p. 24)

Creativity is also included as a component in the prominent theories of giftedness (Kaufman, Plucker, and Russell, 2012). The theory of Renzulli (2005), for example, includes giftedness as emerging from the interaction of three components; above average ability, creativity, and task commitment. The National Association of Gifted Children (NAGC) (2010) published a position paper proposing a definition of giftedness that emphasizes talent development as a lifelong process. Subotnik, Olszewski-Kubilius, and Worrell (2012) indicated that "giftedness can be viewed as developmental in the beginning stages in which potential is the key variable" (p. 176). Giftedness, however, has been used interchangeably with the concept of talent, and both are defined in different ways with no one definition that is universally accepted (Davis, Riman, & Seigle, 2011).

In turn, mathematical giftedness and talent can be demonstrated in a variety of ways. Although academic achievement in mathematics is a strong predictor of mathematical talent, the absence of it does not necessarily mean the absence of potentials. Students are different in terms of being stimulated to express their academic abilities, and "absence of evidence is not evidence of absence". Sheffield (1994) stated that "the frequent narrow definition of gifted mathematics student is scoring above the 95th percentile on a test of mathematical achievement" (p. 3). The NCTM defines the group of students with high ability in mathematics as mathematically promising; the NCTM Task Force on Mathematically Promising Students identifies mathematical promise as "a function of ability, motivation, belief, and experience or opportunity with large range of abilities and a continuum of needs that should be met" (Sheffield, 1999, p. 310).

As a result of this ambiguity and overlap between many concepts including academic achievement, giftedness, talent, intelligence, and creativity, it is difficult to achieve accuracy in the first critical process to provide gifted students with special enrichment programs, e.g. in identification processes. The identification processes of gifted students for special educational services do not only lack clear definitions of giftedness and talent, but also are affected by many other factors such as psychological and cultural factors. Therefore, the likelihood that some gifted students are excluded from special pull-out educational services does exist (Davis, Rimm, & Siegle, 2011; Wadaani, 2015a).

Therefore, general classrooms in many schools may include gifted students who are not identified as gifted and are deprived of additional special enrichment services because of an inaccurate identification process, or because their schools do not offer special programs for gifted and talented students. Based upon that, and taking into account the need of nurturing student creativity, and the importance of supporting gifted students in general classrooms, teachers should internalize broad conceptions of creativity and talent, work to foster creativity skills for all students, and provide additional enrichment opportunities for those who have high abilities in general classrooms (Wadaani, 2015a).

Burleson (2005) indicated that notable psychologists agree that learning is enhanced when it is pursued as a creative and self-actualizing passion. Treffinger, Young, Selby, and Shepardson (2002) added that deliberate efforts to nurture creative thinking skills are important components of excellent educational programs. In mathematics education field, creativity is considered the essence that should not be neglected in general classrooms in order to facilitate the development of young mathematicians (Leikin & Pitta-Pantazi, 2013; Mann, 2006). Encouraging creativity in mathematics classes is important for all students to enjoy working in mathematics and to develop a meaningful understanding of mathematical concepts. It is more important for students who have high abilities in order to develop their mathematical talent and to become more creative and future mathematicians.

However, nurturing creativity and supporting mathematical talent development in general education schools require that teachers hold a clear understanding of these multifaceted interrelated concepts, hold positive attitudes towards gifted education, and have facilities and support to teach for creativity and mathematical talent development. Following is a proposed developmental model for further clarification of such concepts and phases regarding creativity and mathematical talent development; teaching for creativity and teaching for mathematical talent development are included as the two major parts of the model that can be utilized for a sequence of student development in mathematical proficiency: gifted, talented, and mathematicians. This model is based on a conceptual framework that includes creativity in mathematics as a malleable multi-level ability that is considered the essence of mathematical learning and accomplishment, with the assumption that all students have the potential of expressing and developing mathematical creativity, and some of them may further advance in mathematics to a level of mathematical talent development that results in original impacts in the domain.

6.6 RESULTS AND DISCUSSION

T-CMTD is a recommended model for supporting creativity and mathematical talent development in general education schools based on clear understanding and positive beliefs about student capability of success. The T-CMTD model requires that teachers hold positive attitudes towards effective teaching as a process of human development. Teachers who hold positive beliefs and passion for effective teaching will look for professional development opportunities, so he/she can use different teaching methods and provide additional options that nurture creativity and mathematical talent based on students' needs and the situation of each educational setting.

In order to teach for creativity and mathematical talent development, teachers should understand their roles as a facilitator of optimal human development, conceptualize creativity as an ongoing process of self-development, and hold positive beliefs about student capability for success. The mission is to provide opportunities for all students to develop academic, personal and social skills necessary to become successful individuals, increase their achievement to reach the highest level possible, become gifted, then talented, and perhaps become future mathematicians.

Adopting T-CMTD as a model of teaching entails that creativity is perceived as "Human Development toward Self-Actualization". Self-actualization was defined by Maslow (1943) as "the desire for self-fulfillment and being everything that one is capable of becoming" (p. 382). The humanistic self-actualization approach to creativity (Maslow, 1943, 1968; Rogers, 1954) has been supported either explicitly or implicitly by many scholars. According to Davis (2004), the relationship between creativity and self-actualization is one of the most influential concepts in the field of creativity, and argued that creativity is a way of thinking and living that leads to personal development and a productive successful life. Creativity, as human development, includes a broad range of personality characteristics, life skills, general mental skills, and domain-related skills (specialized creativity).

However, mathematically gifted students could be viewed as individuals who have above average aptitudes in mathematics with average general intelligence, and task commitment with strong motivation and attitudes towards continued learning and applications of mathematics in real-life situations (creativity: personality). The giftedness with above average general intelligence can turn to become talent if creativity (creativity: person and process) is developed as a major component of talent development. Original productivity in the discipline (creativity: person, process, and product) requires not only talent but also more time, task commitment, and further environmental support for creativity (creativity: person, process, product, and press/environment). Teachers should help each student through the T-CMTD model to reach the highest level possible of overall development including maximizing academic achievement, nurturing creativity, and mathematical talent along with personal and social skills necessary for authentic growth and life success.

Teaching for creativity in this case means facilitating the ongoing overall development of humans/students that leads them to positively actualize themselves and reach the highest level possible appropriate for their special aptitudes, potentials, circumstances, and community needs. As the highest level of self-actualization may vary from an individual to another, students should be facilitated to develop positive belief systems related to their personal abilities, achievement goals, and future success, and lead them to keep an active and regenerated desire for self-actualization through unlimited high-level personal goals of achievement. Based on the T-CMTD model, teachers should teach creativity to all students considering each student's unique aptitudes, circumstances, as well as the educational situation. Teaching for creativity requires certain classroom conditions and teaching practices, as discussed in the literature review, that provide a healthy environment for all students to express and develop their creativity and mathematical abilities. Figure 6.1 illustrates the researcher's view of the T-CMTD model.

Creativity in Phase I of this model can be viewed as general learning, personal, and social skills; it might be mathematical creativity at a low level. All students benefit from the creativity atmosphere in Phase I, move forward in their meaningful effective learning and personal/social development, and may express giftedness in mathematics later. It is highly possible that some students are encouraged to express high ability in mathematics/high achievement/giftedness, and it is also possible that other students are influenced and enhanced by the creativity environment to systematically work hard to achieve self-actualization, and may extend their abilities and become potentially gifted. Students who have high potential in mathematics should be provided with additional enrichment opportunities in Phase II.

In Phase II, gifted students are facilitated to further develop their potential mathematical talent and creativity in mathematics, for being future mathematicians who can develop original produces with time, if intrinsic motivation/task commitment and environmental support for creativity continue to be available. The focus in Phase II is on mathematical creativity development, but it does not mean neglecting the development of general creativity as skills of effective learning, personal, and social growth, which should be an ongoing development process for all students as the essence of the philosophy of teaching for creativity and mathematical development. In Phase II, possible services for the gifted students include special activities and programs inside or outside the classroom that can be extended for further enrichment out-of-school environment. The aim of this model is to develop not only mathematical talent but also the necessary skills for all students to be effective creative continued learners which will lead them to high level of self-actualization, and perhaps

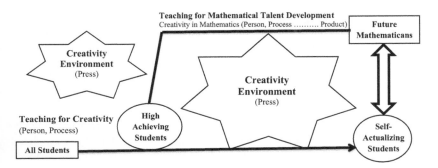

FIGURE 6.1 Model of teaching for creativity and mathematical talent development T-CMTD.

some of them moving further ahead, with enrichment programs, to be mathematicians with needed personal and social skills that make their contributions in the field and their society more effective and valuable.

Effective teaching practices depend on several factors related to the components of each educational situation. Best decisions related to teaching methods used are expected from teachers themselves; teachers, as classroom experts, should understand more than others what might effectively help their students achieve the desired goals in each educational setting. Teaching practices for developing creativity and mathematical talent might include, but are not limited to:

- ensuring that students have prior knowledge and skills needed to succeed in the new topic.
- using advanced organizers and inquiry activities to help students construct knowledge and skills.
- providing needed learning tools with utilization of technology.
- allowing independent thinking.
- encouraging multiple methods and responses.
- informing students of the creative process and creative individuals.
- developing creativity as personal/social/life skills.
- developing creativity as a process/thinking skills.
- assigning heterogeneous group tasks.
- helping students feel safe and enjoy learning.
- encouraging social communication/sharing.
- respectfully discussing and evaluating of ideas.
- effectively using questions.
- providing positive feedback.
- building upon student ideas.
- wisely assigning group tasks.
- differentiating of instruction and evaluation.
- helping students feel of the aesthetic and benefits of mathematics using real-life applications.
- collaborating with significant others to provide additional remedy or enrichment activities such as tutoring and university level advanced experiences.

The T-CMTD model might lead students to move through different developmental stages of academic and life success. This model is based on the researcher's view, in light of theoretical literature, about the characteristics of the individuals in each level of mathematical proficiency development for the purpose of understanding the T-CMTD. model. Although this model might be more applicable in the field of mathematics, especially that general IQ is included as a criterion, it does not propose a generalizable sequence of human development in mathematics or other fields as the researcher believes that human intelligence is an elusive phenomenon. Creativity in this model is an essential component that exists at each level of mathematical ability development; however, it has different elements at each level as creativity can be nurtured first as personality skills, then can be developed into personality skills and skills of thinking/processes, and later, after considerable time, continued environment support, hard work, and incubation, illumination may appear and novel ideas or products are released for verification; creativity in this case represents personality, process, press, and product. Figure 6.2 simplifies the possible transitional stages of mathematical talent development through the T-CMTD model.

The T-CMTD model is designed to have an open gate for any student to become gifted in mathematics and to develop their potential for being a talented and future mathematician if certain conditions are met with continued availability and ongoing utilization. These conditions are related to environment, personality, and time. Although it is rare that students in K-12 schools highly accomplish in mathematics to the level of mathematicians, they still should be facilitated towards achieving this aim at a future time with positive beliefs about student capability for success. Accordingly, the T-CMTD model is recommended to be utilized for establishing the optimal school environment for facilitating students to move towards the highest level possible of these developmental stages in mathematical accomplishment.

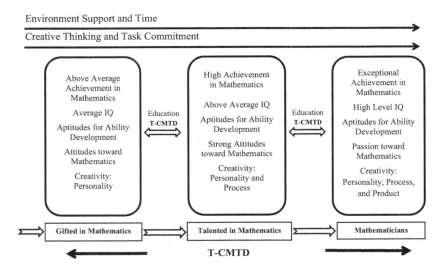

FIGURE 6.2 Potential stages of creativity and mathematical talent development through T-CMTD.

However, the focus group interview with a sample of mathematics teachers revealed multiple types of requirements for the effective utilization of this model in actual environments of general education schools. The qualitative data were transcribed and organized to be around several key aspects including teachers, classrooms settings, school leadership, and curriculum objectives. The study sample of teachers showed support to the proposed T-CMTD model and attitudes to get benefits from its elements, and they emphasized some major needs that are essential for them to be able to apply the model in the best possible way. They indicated the need for effective professional development in the field of creativity and gifted education at school-district training level, and at university program level including more preparation for this type of model.

Teachers also mentioned that the effective application of the T-CMTD model requires some changes in school policies and classrooms set-up, where they pointed out requirements of fewer classroom teaching hours per week (15 maximum), with fewer students in each classroom (20 maximum). Based on teachers' perspectives, these modifications would allow time space for them to plan and well prepare for using such effective teaching models. Other requirements for the effective application of the proposed T-CMTD model discussed by the sample of this study included improving school facilities such as science and mathematics lab, reconsidering the appointment process of school leadership team members, and providing appropriate education for them to become more supportive to teachers in using a new effective model of teaching. Teachers who participated in this study also argued for learning objectives that are emphasized in the current curriculum as they indicated the need for shifting the focus towards life direct learning objectives that reflect on student authentic development.

Overall, the proposed T-CMTD model seemed to be clear and attractive for teacher to adopt; however, current school environment might not be fully ready to include this model, as teachers in this study indicated multiple basic requirements for its effective application. Therefore, educational policymakers and stakeholders are invited to consider such needed development in education systems that allows using new proposed promising teaching models for human development.

6.7 CONCLUSION

Literature includes multiple perspectives about creativity and giftedness, but the dominating perspective is the one that is based on the broad liberal conceptualization of creativity and giftedness with positive beliefs about student capability of success and development. Accordingly, school environment and teachers' attitudes are supposed to be built according to this comprehensive developmental view, with the belief of education as a process of human development. Based on that, this research study synthesized supportive literature for a proposed model that aims at helping teachers get a deep understanding of creativity and mathematical talent development in school settings; the T-CMTD model. Teaching practices are recommended to be aligned with a clear understanding of creativity and talent development as illustrated in this proposed model. Major requirements for the best utilization of this model were discussed to include relevant teacher education and training, less teaching load, effective facilities and appropriate classroom size, school leadership support, and curriculum modification. It is also recommended that researchers make an additional critical review of the proposed T-CMTD model and conduct multiple pilot studies to obtain a scientifically better version that can be utilized in different circumstances of school environment.

REFERENCES

Bahar, K., & Maker, J. (2011). Exploring the Relationship between Mathematical Creativity and Mathematical Achievement. *Asia-Pacific Journal of Gifted and Talented Education, 3*(1), 33–48. Retrieved from http://www.apfgifted.org

Beghetto, R., & Kaufman, J. (2007). Toward a Broader Conception of Creativity: A Case for "mini-c" Creativity. *Psychology of Aesthetics, Creativity, and the Arts, 1*(2), 73–79. doi: org/10.1037/1931–3896.1.2.73

Brinkman, D. (2010). Teaching Creatively and Teaching for Creativity. *Arts Education Policy Review, 111*, 48–50. doi:10.1080/10632910903455785

Burleson, W. (2005). Developing Creativity, Motivation, and Self-actualization with Learning Systems. *International Journal of Human-Computer Studies, 63*, 436–451. doi:10.1016/j.ijhcs.2005.04.007

Davis, G. (2004). *Creativity Is For Ever.* Dubuque, IA: Kendall Hunt Publishing Company.

Davis, G., Rimm, S., & Siegle, D. (2011). *Education of the Gifted and Talented.* Boston, MA: Pearson Education.

Gardner, H. (1983). *Frames of Mind: The Theory of Multiple Intelligences.* New York: Basic Books.

Kaufman, J., Plucker, J., & Russell, C. (2012). Identifying and Assessing Creativity as a Component of Giftedness. *Journal of Psychoeducational Assessment, 30*(60), 60–73. doi:10.1177/0734282911428196

Lee, K., Hwang, D., & Seo, J. (2003). A Development of the Test for Mathematical Creative Problem Solving Ability. *Journal of the Korea Society of Mathematical Education, 7*(3), 163–189. Retrieved from http://icms.kaist.ac.kr

Leikin, R. (2009). Bridging Research and Theory in Mathematics Education with Research and Theory in Creativity and Giftedness. In R. Leikin, A. Berman, & B. Koichu (Eds.), *Creativity in Mathematics and the Education of Gifted Students* (pp. 385–411). Rotterdam, The Netherlands: Sense Publishers.

Leikin, R., & Pitta-Pantazi, D. (2013). Creativity and Mathematics Education: The State of the Art. *The International Journal of Mathematics Education ZDM, 45*,159–166. doi:10.1007/s11858-012-0459-1

Mann, E. (2005). Mathematical Creativity and School Mathematics: Indicators of Mathematical Creativity in Middle School Students. Doctoral Dissertation, The University of Connecticut. Retrieved from http://www.gifted.uconn.edu

Mann, E. (2006). Creativity: The Essence of Mathematics. *Journal for the Education of the Gifted, 30*(2), 236–260. doi:10.4219/jeg-2006-264.

Maslow, A. (1943). A Theory of Human Motivation. *Psychological Review, 50*(4), 370–396. doi:10.1037/h0054346

Maslow, A. (1968). *Toward a Psychology of Being.* New York: Harper.

Miligram, R., & Hong, E. (2009). Talent Loss in Mathematics: Causes and Solutions. In R. Leikin, A. Berman, & B. Koichu (Eds.), *Creativity in Mathematics and the Education of Gifted Students* (pp. 149–161). Rotterdam, The Netherlands: Sense Publishers.

National Association of Gifted Children (NAGC). (2010). *Redefining Giftedness for a New Century: Shifting the Paradigm.* Washington, DC: NAGC. Retrieved from http://www.nagc.org

National Council of Teachers of Mathematics (NCTM). (1980). *An Agenda for Action: Recommendations for School Mathematics for the 1980s.* Reston, VA: NCTM.

National Council of Teachers of Mathematics (NCTM). (2000). *Principles and Standards for School Mathematics.* Reston, VA: NCTM.

Plucker, J., Beghetto, R., & Dow, G. (2004). Why Isn't Creativity More Important to Educational Psychologists? Potential, Pitfalls, and Future Directions in Creativity Research. *Educational Psychologist, 39*, 83–96. Retrieved from http://eric.ed.gov

Renzulli, J. (2005). *The Three-Ring Definition of Giftedness: A Developmental Model for Promoting Creative Productivity.* Storrs: The Naeg Center of Gifted Education and Talent Development, University of Connecticut. Retrieved from http://www.gifted.uconn.edu

Rhodes, M. (1961). An Analysis of Creativity. *Phi Delta Kappan, 42*, 305–310. Retrieved from http://www.jstor.org

Rogers, C. (1954). Toward a Theory of Creativity. *ETC: A Review of General Semantics, 11*, 249–260. doi: 1955–05283-001.

Sheffield, L. (1994). *The Development of Gifted and Talented Mathematics Students and the National Council of Teachers of Mathematics Standards.* Storrs: The National Research Center on the Gifted and Talented, The University of Connecticut.

Sheffield, L. (1999). Serving the Needs of the Mathematically Promising. In L. Sheffield (Ed.), *Developing Mathematically Promising Students* (pp. 43–55). Reston, VA: NCTM.

Sheffield, L. (2013). Creativity and School Mathematics: Some Modest Observations. *The International Journal of Mathematics Education ZDM, 45*, 325–332. doi:10.1007/s11858-013-0484-8

Silver, E. (1997). Fostering Creativity through Instruction Rich in Mathematical Problem Solving and Problem Posing. *The International Journal of Mathematics Education ZDM, 3*, 75–80. doi:10.1007/s11858-997-0003-x

Sriraman, B. (2005). Are Giftedness and Creativity Synonyms in Mathematics? *The Journal of Secondary Gifted Education, 17*(1), 20–36. http://files.eric.ed.gov/fulltext/EJ746043.pdf

Sriraman, B., Yaftian, N., & Lee, K. (2011). Mathematical Creativity and Mathematics Education: A Derivative of Existing Research. In B. Sriraman, & K. Lee (Eds.), *The Elements of Creativity and Giftedness in Mathematics* (pp. 119–130). Rotterdam, The Netherlands: Sense Publishers.

Stanley, J. (1991). An academic Model for Educating the Mathematically Talented. *Gifted Child Quarterly, 35*(1), 36–42. doi:10.1177/001698629103500105

Sternberg, R. (1985). *Beyond IQ: A Triarchic Theory of Human Intelligence*. New York: Cambridge University Press.

Sternberg, R. (2010). Teaching for Creativity. In R. Beghetto, & J. Kaufman (Eds.), *Nurturing Creativity in the Classroom* (pp. 394–414). New York: Cambridge University Press.

Sternberg, R., Kaufman, J., & Grigorenko, E. (2008). *Applied Intelligence*. New York: Cambridge University Press.

Subotnik, R., Olszewski-Kubilius, P., & Worrell, F. (2012). A Proposed Direction Forward for Gifted Education based on Psychological Science. *Gifted Child Quarterly, 56*(4), 176–188. doi:10.1177/0016986212456079

Torrance, E. P. (1995). Insights about Creativity: Questioned, Rejected, Ridiculed, Ignored. *Educational Psychology Review, 7*(3), 313–322. doi:10.1007/BF02213376

Treffinger, D., Young, G., Selby, E., & Shepardson, C. (2002). *Assessing Creativity: A Guide for Educators*. Storrs, CT: The National Research Center on the Gifted and Talented. Retrieved from http://eric.ed.gov

Wadaani, M. R. (2015a). *Teachers' Attitudes and Features of Support Related to Teaching for Creativity and Mathematical Talent Development in the United States*. Lawrence, KS: University of Kansas, Doctoral Dissertation.

Wadaani, M. R. (2015b). Teaching for Creativity as Human Development toward Self-actualization. *Creative Education, 6*, 669–679. doi:10.4236/ce.2015.67067

Wadaani, M. R., Seidl, M., Mends, D., Lagat, A., Edgar, K., Degnan, K., Kunnen, E. (2016). *Education and Human Development*. New York: Magnum Publishing LLC.

Yuan, X., & Sriraman, B. (2011). An Exploratory Study of Relationships Between Students' Creativity and Mathematical Problem-posing Abilities: Comparing Chinese and U.S Students. In B. Sriraman, & K. Lee (Eds.), *The Elements of Creativity and Giftedness in Mathematics* (pp. 5–28). Rotterdam, The Netherlands: Sense Publishers.

For Product Safety Concerns and Information please contact our
EU representative GPSR@taylorandfrancis.com Taylor & Francis
Verlag GmbH, Kaufingerstraße 24, 80331 München, Germany